Angela Locke has been writing seriously for about ten years and has published one previous book about Cumbria. Born in Suffolk, she trained as a teacher and taught in various parts of Britain before her marriage, also working as a freelance journalist. A confirmed country-lover, she has a special affection for the sea and high places. She and her husband live in Sussex with their three children and an army of dogs, cats and other assorted animals. She spends as much time as possible wandering in the woods around their house.

To
with all g—
wishes

Angela Locke.

SEARCH DOG

by

Angela Locke

SPHERE BOOKS LIMITED

SPHERE BOOKS LTD

Published by the Penguin Group
27 Wrights Lane, London W8 5TZ, England
Viking Penguin Inc., 40 West 23rd Street, New York, New York 10010, USA
Penguin Books Australia Ltd, Ringwood, Victoria, Australia
Penguin Books Canada Ltd, 2801 John Street, Markham, Ontario, Canada L3R 1B4
Penguin Books (NZ) Ltd, 182–190 Wairau Road, Auckland 10, New Zealand

Penguin Books Ltd, Registered Offices: Harmondsworth, Middlesex, England

First published in Great Britain in hardback by Souvenir Press Ltd 1987
Published by Sphere Books Ltd 1988

Printed and bound in Great Britain by
Richard Clay Ltd, Bungay, Suffolk

This book is dedicated to SAM, with much love, and to all the handlers and dogs in SARDA who do so much for Search and Rescue in Britain.

The characters in this book, apart from John, Tina and their family, are entirely fictitious. The same is true of the canine characters, with the exception of SAM. Most references to the fells, and to individual teams, have been disguised.

Illustration Credits

Sam as a puppy.   *Courtesy John Brown*

Practising for his misdemeanour at his first Annual Course?   *Courtesy John Brown*

John and Sam begin their training on the crags. *Courtesy John Brown*

Novice Search Dog Sam with the SARDA Shield he won in 1980.   *Courtesy Cumbrian Newspapers Group Ltd*

Search Dog Sam, in his scarlet SARDA coat. *Courtesy John Brown*

Sam finds the 'casualty' and homes in.   *Photo: Paul Fearn*

In his enthusiasm, he attempts to lick the 'casualty's' face. *Photo: Paul Fearn*

Sam 'speaks' to John, so bringing him to the right spot. *Photo: Paul Fearn*

Keswick MRT ambulance, ready with stretchers and other rescue equipment.   *Photo: Paul Fearn*

A Keswick team member checks the map references. *Photo: Angela Locke*

John listens to the 'crack' (the briefing) at Keswick MRT. Behind him are radio sets that will be carried by the handlers. *Photo: Angela Locke*

Sam in his element, surveying his home territory. *Courtesy John Brown*

When landing conditions are impossible, the casualty has to be winched up. Sea King helicopters, used for longer-range operations and search work, from 202 Squadron, RAF Boulmer.   *Courtesy John Brown*

Using smoke to indicate wind direction for the helicopter coming in to pick up the casualty — Langdale/Ambleside team.   *Courtesy John Brown*

The Sea King helicopter lands safely. John and Sam wait in the foreground.   *Courtesy John Brown*

Young Tyan picks up tips from Sam on a Search exercise. *Courtesy John Brown*

Collies and handlers prepare for a Search exercise. *Courtesy Roly Grayson*

Sam with Tyan — full Search Dog with a hopeful trainee. *Courtesy John Brown*

Still very much a family dog: Sam with Tina, the children Matthew, Anna and Pippa, and Tyan.   *Courtesy John Brown*

PREFACE

It is only a little over twenty years since the idea of SARDA was born, and much less than that since the awareness that 'search dogs' exist in Britain has begun to filter through to the public consciousness. And yet, for many weeks of the year, these dogs are out in the wild places of the British Isles, occasionally twice or three times in one day, looking for lost people. One day it might be us.

For many casualties, the first time they ever meet a Search Dog is that joyful moment when, after long hours lost on a fellside, they come face to face with a barking animal who appears from nowhere out of the mist and snow. Is he friendly? Has he got a human being with him? He disappears again, perhaps for many minutes; he may appear and disappear several times more. He may even take away an article of the astonished casualty's possessions. Very gradually it may dawn on the befuddled victim that, incredibly, the dog has found him by himself, and is leading the rescuers in. For such a person, perhaps in a life and death situation, SARDA dogs have become a warm, welcome reality.

Recently two SARDA dogs, with their handlers David Riley and David Jones, were flown out to El Salvador to help search for earthquake victims after the disaster there. Their presence meant that at least one life was saved. They were echoing the first formal use of Search Dogs in this country more than forty years ago, when dogs were used during the London blitz to locate victims of the bombing. They, too, used air scent and saved many lives. One of them, an Alsatian, even won the Dickon Medal for exceptional bravery under conditions of extreme danger.

Yet it was in the Alps that the first formal ideas of doing just that . . . using air scent to track down avalanche victims or other buried casualties, first became a reality. Dogs had been used since the seventeenth century in Switzerland, to help the monks of the hospice built by Bernard of Menthon to give assistance to travellers across the Great St Bernard Pass. Initially they were used as guides in bad conditions, but soon they had graduated to indicating to the monks where someone might be found, even under feet of snow.

But the real beginnings of the avalanche Search Dog organisations as they exist today came with the formation of the Swiss Alpine Club after the last war. As far back as the First World War dogs had been used by the Red Cross to locate injured men on the battlefields during lulls in the fighting. These dogs were trained to find by airborne scent. Then came a moment in the winter of 1937–38 during an avalanche search. A team member's dog, which had been accompanying the search, began to show interest in one particular place and eventually began to bark. The spot had already been probed, but after reprobing the victim was found, still alive.

This incident led a Swiss dog training expert to train four Alsatians to search for avalanche victims and these were then given to the Swiss Army. They proved very successful and, after the war, the Swiss Alpine Club decided to train dogs for avalanche work for their own rescue network. At the present time avalanche dog training centres have become widespread throughout the Alps.

But it needed one man to bring the whole idea of Search Dogs back to Britain where, since the war, the usefulness of dogs as searchers had almost been forgotten. Hamish MacInnes, a foremost authority on Mountain Rescue in Britain, was in 1963 the leader of Glencoe Mountain

Rescue Team. He was very interested in the idea of using dogs in a search capacity and was invited by the International Red Cross to attend a training course for avalanche dogs at Engelbert in Switzerland.

What he learned there convinced him that there was a valuable rôle for Search Dogs to work with our own Mountain Rescue Teams. The problems in Britain were not the same. A missing person does not necessarily present the same difficulties as an avalanche victim, and for dogs in Britain most of their work would be concerned with searching over a large area, often for long periods of time. The dogs would need to be more versatile, and must be capable of sustaining their interest and commitment for hours and days if necessary. It was a tall order.

On his return from Switzerland, Hamish MacInnes trained his own Alsatians, Rangi and Tiki, with such success that in December 1964 he held a pilot course of his own at Glencoe to which he invited six handlers and their dogs. The results were so encouraging that in May 1965 he held a meeting at his home which resulted in the formation of the Search and Rescue Dog Association. It was decided that an Annual Course would be held every year in December at Glencoe. Dogs would attend for training and assessment. Thus SARDA was born, and at that first Annual Course dogs from all over Scotland, England and Wales were present for the inaugural meeting.

Since 1971 there have been three separate Associations —one for each country. Wales and England tend to train together, but Scotland, having its own particular problems of snow searching and avalanches, holds its own courses, and liaises with the other two. For the first time, there is now a graded Search Dog attached to the Mountains of Mourne Mountain Rescue Team in Ireland, and other dogs are in training there.

I would never have known any of this if my sister had not had the good fortune to live next door to the owner of Sam.

9

That was how it all began. Before that I knew nothing about SARDA. But from the moment I saw the sign SEARCH DOG in the back of John's car and, having knocked tentatively on the door, found myself overwhelmed by a large yellow Labrador which promptly laid his heavier half on my lap and went to sleep, I was hooked.

I have to say that my next arrival in Cumbria, this time to observe the November course at Hammerbank, was filled with dread. I arrived at Carlisle station in the rain with no one to meet me, and a rucksack stuffed with what I afterwards discovered was hopelessly inadequate clothing. The thought flashed through my mind that, although this was the second book I had written about the Lake District, my next might profitably be set in the Bahamas . . .

But then, once again, I was hooked. To watch Sam in action at two thousand feet, chomping his jaws for dear life, tail wagging, was an unforgettable experience. Until then I don't suppose I had really believed that a dog could find a 'body' . . . there, in appalling conditions, cunningly hidden on a vast fellside, with the impenetrable 'clag' making every step hazardous.

That moment convinced me that I must write this book. It saw me through many things. Penrith team, who did so much to help me with the mechanics of Mountain Rescue, persuaded me to abseil over a (smallish) precipice; John got me a couple of thousand feet up the face of Helvellyn with a wind-chill factor of –30C. There was a weak moment when I began slipping cramponless back down the mountain with Sam rifling callously through my pockets for Mars bars. Thanks, Sam. I don't think I would have got up if John had not threatened to switch on the radio so that everyone could hear my complaints!

And Penrith enabled me to take my long-suffering brother-in-law Paul Fearn up to the roof of the Pennines in the worst conditions for years to photograph an exercise. This led to the immortal moment which my youngest daughter enshrined in a letter to her sister . . . 'my mum got hyperfiurmia (sic) and was raped in a plastic bag.' The truth, always more mundane, involved being wrapped in a polybag and lying in a bog, which was a jolly sight colder than moving around. But most sincere thanks to Ken Thompson for evacuating me off the moor, soaked and very cold, and for giving me his last corned beef sandwich. Admiration, too, for the BBC camera team who were attempting to film the exercise, and who had humped over a hundred pounds of sound and video equipment through miles of bog, refusing to be evacuated despite pleas from the Controller. True professionalism.

It is only when you are actually there that you can understand how easy it is to make mistakes . . . to get lost on a barren, featureless moor, to fall into a bog and get soaked and, stumbling through the tussocks of sodden heather, to begin to experience the strangely disorientating feeling which signals the onset of hypothermia.

But I would not have missed it for the world. To see the dogs in action; to experience the 'buzz' of a successful find; to climb and climb, until your leg muscles scream for rest; to stop and look back, to see the grey lake spread far below. The warmth of Sam, and his instinctive friendship and trust, and the constantly joking, teasing companionship of those committed men and women who make up the Mountain Rescue Teams . . .

As one of them said to me, 'If you're carrying a stretcher down the mountain and the head of the "foxtrot"* keeps falling off the body, what else can you do to keep sane but have a crack and a joke . . . ?' That is the serious side. Underneath, they are quiet people who shun publicity, whom you feel you could trust with your life. They are inspirational.

MRT team members and SARDA handlers do not seek

---

*MRT code for a dead body.

personal publicity. They see there is a job to be done; they do not make a fuss about it. They are members of a team. One of them even objected strongly to having his photograph taken. It is partly for this reason that most references in this book to the fells, and to individual teams, have been disguised. The second reason is even more important. Sometimes, the stories deal with individual tragedies. Always there is pain and sometimes embarrassment involved with anyone who has been lost or injured in the fells. Thus, although I have drawn on many individual incidents, I have, for the sake of the casualties, and also out of respect for relatives of any fatalities, tried to disguise any sources as effectively as possible, and to use only fictional characters, apart from John and his family.

I have taken part, in a limited capacity, in three search and rescue operations—two successful and one which had a tragic ending. I have listened to innumerable stories. I have ploughed through endless case histories. From these, I have tried to construct a vivid picture of the life of a Search Dog. I hope I have succeeded. Sam's character is real. His background is real. The stories must be taken as an amalgam of all Search Dog incidents, and must not be related to individual case histories.

I have to admit to bias. Sam is a wonderful dog. He is an incorrigible character—dedicated, brilliant, but given to embarrassing his owner with his propensity for 'the ladies', as my mother primly put it, and for fighting with rivals, neither of which he will probably grow out of until he draws his last breath. Last November I watched, fascinated, on the side of the Pennines, as he fixed his deadly rival with a ferocious glare during a TV interview and, in a moment of less then perfect concentration on John's part, pinned the poor Border Collie (his long-term enemy) to the ground in front of the camera.

I, too, have experienced his particular leanings; once, after a visit to my sister, I stopped at John's house to check

12

on a few final points before our next meeting. When I returned to the car, I found Sam snugly ensconced next to my Golden Retriever bitch, who looked disgustingly happy about the whole thing. Love at first sight. He was obviously ready to travel, if necessary, south to Sussex, in order to be in her company. Sam may be a bit of a star, like all the SARDA dogs, but it doesn't stop him getting up to mischief. As one of the team members put it to me, 'Ay, he were a rascal when he were young.' Given half a chance I think he would still be a rascal now . . . that's his charm.

There are many heroes and heroines among the dogs, and among their handlers, too. I cannot realistically write about all of them. All I can try to do is pay tribute to them.

SARDA is a registered charity. After working with them for so long, I know how much they struggle with inadequate funds and how worthy a cause it is. I hope that this book will not only entertain but will also help towards an awareness of what those handlers and dogs, who turn out every night in sometimes nightmare conditions, actually do.

I must also stress that some techniques of training and assessment have changed since Sam began his career. Police Dog Handler Neville Sharp, for example, now conducts stringent obedience tests at the very onset of training, and many training structures have been revised. SARDA is a young organisation, which is growing all the time. There will undoubtedly be many more changes as the years progress. The essential principles, however, remain the same.

I have to give specific thanks to a number of people. First and foremost must be John Brown, without whom this book would never have come to fruition. His help, whether with anecdotes and information, or with the provision of marvellous photographs, both from his family album and, more recently, specifically for the book, has been given unstintingly, with one thought in mind: the recognition of

13

the work of SARDA. I could not have done it without him, or without Tina, his wife, who so patiently coped with phone calls and late nights and all the disruption, and goes on coping. Thank you both. And of course the committee of SARDA who allowed me to attend the training sessions and the assessment courses, as well as other meetings, and who have given me unstinting help and a warm welcome over the time I have been writing the book. In that category I have to mention Pete Durst, Chairman, Malcolm Grindrod, until last year SARDA's training officer, David Riley, the course organiser, and Dave Brown, SARDA secretary. Also in SARDA, Ken Saxby, for his advice about stock training, Ian Cutler, for enlightening me about the trails of a Novice Search Dog, and Kipp, for making me laugh. After attending the 1987 course I must also thank: Neville Sharp, the Police Dog handler; Neil Powell of SARDA Ireland, for advice and for lending me his carry mat so that I could 'body' out on the fell; Alison Colau, for so generously parting with her sleeping bag; Dominic Atkinson, and Flt. Lt. Roly Grayson of the RAF Survival Squadron, who spared time after being a 'body' for seven hours a day to take pictures of a one o'clock in the morning call-out in the middle of the course.

I would also like to thank Penrith MRT, and their team leader Jed Feeney, for giving me so much help in the way of rescue exercise and advice, and even for persuading me to abseil down a cliff face; and Mike Nixon, Martin Bellarby, and the Keswick MRT HQ for allowing me on two occasions to take pictures during call-outs. And I would like to say a very big thank you to my sister, Stephanie Fearn, who probably wishes sincerely that I would stop writing books about Cumbria, stomping snow all over her carpets, and filling the place with wet walking boots. It was she who began it all, by listening to the radio and realising that the dog next door, an unremarkable-looking Labrador, was actually the SEARCH DOG SAM whose 'find' was being reported on Radio Cumbria. Stephanie has been a wonderful help; Paul, her husband, went through hell and high water to take some excellent photographs in terrible conditions up in the Pennines, and later that same day turned out late at night to photograph a surprise call-out at Keswick MRT.

Thank you, too, to Gordon and Suzanne Staniforth for respectively reading my manuscript for accuracy, and helping, together with my long-suffering husband and my mother, to look after my smallest children so warmly and well while I went walkabout. I do not think they missed me at all! And grateful thanks to Carolyn Pollitt for all her secretarial help.

Last of all, thank you, Sam, for being such a fool, and for being so very clever, both at the same time. This book is all for you. Because you are the bravest and the softest dog I know.

Angela Locke
February, 1987

1

There must be a point where you give up, lie down in the snow, and just go to sleep. Ben had been thinking that for a while. This was a Survival course; he was the team leader. It was ridiculous to think that he would ever let go, close his eyes, forget all his responsibilities, just so that he could stop the pain of walking, of yomping for mile after mile through the sodden heather, of facing the wind which smacked him in the face whenever he looked up . . .

He swivelled his body round, feeling the wind buffeting his back, to check on the other two. Carl looked in pretty bad shape, Andy seemed OK, but . . . ? They had stopped, heads bent, hoods up, too tired to smile or speak.

'If we don't get back on the path soon, we'll have to bivvy. Let's have another look at the map,' he said.

Carl slumped down on the wet heather.

'How long now, Ben? I've just about had it.'

Ben handed round a lump of sodden chocolate and fought for a few moments with the equally wet map which threatened to blow away in his hands.

'For God's sake, let's just have a kip.' It was Andy who now, too, was stretched out in the heather, his hood thrown back, eyes closed. His voice sounded slurred. Ben experienced a small spasm of fear.

'We can't sleep here, Andy. You know that. There's no shelter at all. We'll be back on the path soon; it's only a matter of getting the compass bearing right.'

But he thought he had done it *all* right. He had listened to everything they had told him. And when they had set out

17

that morning, at dawn, to make their way to the old shooting hut where they would rendezvous with the other two teams and bivvy for the night, he had been so confident. But if they were on the right track, where was the granite cairn they should have seen hours ago, and where was the beck with the rowan tree? In his heart he knew they must be hopelessly lost.

And now the light was failing. He was finding it harder and harder to see. The misty rain was growing heavier. He thought it might be turning to snow.

'Listen, everyone. We may be lost. If we are, I'm sorry. It's my fault.' He felt as though he was talking to himself. Andy lay stretched out, his face white, his cagoule unzipped, oblivious of the soaking misty rain, and Carl, head bent over his knees, looked all in, not even listening . . .

'We must bivvy for the night. It's stupid to go on walking. We'll just lose body heat . . . get soaked to the skin, then we won't have any reserves left . . .' he struggled to remember the lectures they had had on 'Emergency Procedure' down at the Survival Centre. Then it had seemed a bit of a laugh . . . you never thought it might happen to you. 'Let's just go on a few more yards . . . find a drier spot . . . out of the wind . . .' There was no reply . . .

He had managed to light the primus by the side of a small beck, in a little valley out of the full force of the wind. It was impossible to put up the tent, but he would manage something. The light was almost gone. He had fought against the wind and taken a hot drink back to Carl and managed to rouse him enough to get him to help. Together they dragged Andy's by now unconscious form to the makeshift camp. By the time they had finished, Carl had had it, too.

Ben unpacked the polybags from his rucksack and fed Andy's feet into one of them while Carl, weary and slow, climbed into another. Then Ben wrapped the tent round

18

the two of them. Andy was wet through to the skin . . . he must have been losing body heat for hours. How had he got like that without Ben noticing? He remembered, looking back, that Andy had stepped into a bog a while back—how long ago?—and they had had to haul him out. Andy had complained that the water was icy, and that his feet would not warm up, and afterwards about pains in his back, and lately he had had bouts of shivering . . . how had Ben not taken in the warning signs? He blamed himself. Now they were stuck out on the fell in the winter dusk, hopelessly lost, and the wind was getting up and flurries of snow, which had seemed nothing at first, were building up against the narrow wall of the little valley, on the perimeter of the torchlight, growing fast into drifts.

Had this been a good place to choose?

*   *   *

The call-out had come just as most of them had come in from work and were having their tea, glad to be out of it, this foul weather which hurled itself against the window-panes. For some of them, the journey home had already been a struggle, up narrow fell roads and tracks quickly threatened by snow; they were relieved to be in out of the wet, in front of a warm fire.

The phone calls came, and teas were left and wives or husbands kissed, and in some homes dogs were loaded into the boots of cars, excited dogs with scarlet jackets on their backs; and, within a remarkably few minutes, the teams were meeting at their own rendezvous—sometimes team headquarters—for a first briefing and a longer journey up the skiddy roads onto Stainmoor, to the roof of England, where the fells lay under their first covering of snow.

'. . . three lads overdue for their rendezvous with the team leader, probably went off their route somewhere between here, and here . . .' they clustered round the Land-Rover in the dark, straining to see. 'They are in-experienced but they have had good instruction on survival, and they have all the right equipment . . .

The dogs in their scarlet jackets were allotted their own piece of bleak moorland to search. They set off with their handlers, tails wagging, raring to go—Collies, a couple of Alsatians and, for the first time joining the five Mountain Rescue Teams on the search, a couple of Novice Search Dogs, including a yellow Labrador.

The handlers sent the dogs 'away', and together each dog quartered his own special section of the fell, the handler bent over against the vertical snow, the dog running on, its sensitive nose alert for any hint of the air scent which would indicate a human presence. Meanwhile the laborious line search, made up of Mountain Rescue Team members strung out a few yards apart, struggled across the heather on the far side of the moor. It was a team effort whichever way you looked at it: the only way, in this wild country, of locating three tiny, vulnerable shapes in this vast expanse of white—the dog's nose and its exceptional training, the skill of the handler, and the expertise of the Mountain Rescue Team of which the dog was an integral part. Even so, with all that back-up, it seemed, in such conditions, an impossible task . . .

*   *   *

It was time to go to sleep. It was like being a child again, except that part of Ben knew it wasn't home after all, like when he was little, with the Superman wallpaper and his Airfix at the bottom of the bed. It was really ice-cold moorland with nothing between him and death but a survival pack and the clothes they stood up in. And he was team leader and he mustn't go to sleep. He kept crawling over to the other two, and slapping their faces, like they did

20

in films, because he could no longer remember what it was he was supposed to do . . . but Andy did not even respond any more and Carl merely groaned and snuggled deeper into his anorak. There hardly seemed to be much point and, after all, the snow was getting deeper and was so soft. Somehow, he thought he could be warm if only he could burrow deep into its feather-lightness, where it lay in drifts on the sides of the beck . . . it was the only welcoming thing . . .

* * *

It was long after dark and in their hearts they had all given up hope. It was the first bad snow of the winter and, as if to make up for its long absence, the blizzard raged with unremitting fury. Surely nothing could survive in this? They trudged on, the crackle of radio contact their only link with the outside world. Each man and woman was an island; they saw no lights; they heard no voices, apart from the occasional bursts of radio contact. They were entirely alone.

Suddenly, as the Controller back at base was wondering whether to bring the teams in off the fell before the conditions became impossible, one of the handlers, weary with cold and wet through, faced his dog in the circle of torchlight. The dog was chomping his jaws in a most peculiar way, making a barking noise at the same time. The yellow Labrador was lit with a strange greenish light, his eyes red in the torchlight.

'Show me, Sam, show me!'

The dog was wild with excitement. He disappeared once again into the white, wind-driven dark. Then he was there again, four-square planted, his paws digging the snow, telling his master. Together they fought across the moor, the Labrador running ahead into the blackness, that strange greenish glow on his back.

There, half-buried in the snow, wrapped in the fabric of a tent, were three small shapes. They were very still. The dog had stopped barking and was nosing at them, whining softly.

'Good dog, Sam! Good dog!'

The handler made a fuss of him and then knelt down beside the bodies, pressing the button of his radio as he did so. Another dog appeared out of the driving snow, barking its head off . . . the SARDA dogs had notched up another find . . .

To read the Incident Report a few months later, you would not have known what it had cost . . . hours of searching the fell in a white-out, lives risked, the white faces waiting on the fell road when the body of a young boy was brought down, the still, white figure in intensive care— Carl fighting for life. Ben alive, also in a hospital bed, never forgetting it was all his fault. Dogs whacked out in front of the fire, maybe two lives saved, wait and see how the boy does; but one lost: can't grieve, otherwise you would never go up again . . . it's a savage country . . . what did someone say? 'If this is God's Own Country He should be punished for it . . .' but maybe He gave us something back . . .

And how had it all begun, that dog, that handler? It had been a long, long journey, and there was still a long way to go. Who would have guessed, in the beginning, how it would all turn out. . . ?

2

'I'm sorry. They've all gone except this one.' The woman smiled down at the puppy, who was lying on his back, scrabbling at her boot.

'To be honest,' she went on, 'he's not really one of our successes. A lot of our customers want to show, and he's got a lot of faults. You can see that.'

John bent down and stroked the pink tummy, engorged with its recent feed. Tiny needle teeth made contact with his fingers, the puppy tying himself in knots as he tried to get his paws in on the act as well.

'He looks OK to me,' he said.

The breeder shook her head.

'You won't get any sales patter from us. We have a reputation to keep up. He's too short in the leg, for a start . . .'

She knelt down to flick the golden puppy over onto his feet. He scampered off, wagging his haunches.

'There!' she said. 'See what I mean? And his head . . . even at this age we can tell. I can't imagine how we came to breed such a puppy. And of course, he's the worst of the lot for getting into mischief, despite his sweet nature. Pulled the side panel off the stable door with his teeth this morning, and the head off my grand-daughter's doll only yesterday.'

'We would like to have him,' John said. He looked across at Tina, who smiled back at him.

The woman looked doubtful.

'Are you sure you wouldn't be better with a dog who has

a quieter temperament? You say you've never had a dog before . . .'

'I had dogs when I was a child. so did John,' Tina interjected.

'And you're living in a flat at the moment,' the woman went on, 'right in the middle of a town. Not a good place to bring up a dog.'

'We're moving to our first house very soon,' John explained, 'And we wondered . . . if you could hang on to him for a few days while we get him a basket, that sort of thing. We could pay for him now . . .' He felt in his jacket for his chequebook.

The woman bit her lip.

'Well, I don't know. I must be honest, we have become rather fond of him ourselves, and my daughter thought he would be good company when she's on her own . . . she would never show him, of course . . . I'll tell you what! My other bitch is due to whelp next week. I could give you pick of litter . . .'

John squatted down on his heels and called to the puppy who, oblivious to all discussions about his future, was digging a large hole in the grass verge, his tail vibrating in the air with excitement.

'Here . . . Come here . . .' The puppy ignored him. 'Has he got a name?'

The breeder laughed. 'I can see I'm on a losing wicket! You really do want him, don't you?'

John looked up at Tina, who smiled down at him.

'Yes, we really do want him, although we appreciate your being so honest with us. I would like to call him Sam, after the yellow Labrador I had when I was a child. He was a real family dog, and a great character. Even though we had other dogs after Sam. I never forgot him.' He paused. 'OK. Let's see if he likes it.'

He opened his arms wide.

'Here, Sam. Here, Sam . . .' The puppy took no notice.

He whistled softly, calling again. At the sound, a small pale head appeared out of the shell crater of what had once been a grass verge, its muzzle speckled with dark earth. The puppy regarded John quizzically, his head on one side. Then, his mind made up, he galloped crabwise across the drive, his over-large paws skittering in all directions. He got within a yard or two of them and fell over on his back, his hind legs spread in an attitude of supplication.

They all laughed.

'I suppose he wants his tummy tickled again,' Tina said.

The breeder nodded.

'One of his many weaknesses. And you'll have a friend for life!'

They both bent down, stroking the soft, downy coat. The puppy closed his eyes in bliss.

'Sam!' John said softly.

The puppy opened one eye, sneezed violently twice, grabbed Tina's scarf with his teeth, and once more galloped off, haunches swaggering with pride at his trophy.

'Oh, no!' Tina exclaimed. 'My best scarf!'

'I did warn you.' The breeder laughed. 'Nothing's safe. That's true of all puppies, of course, but it's especially true of Sam.'

* * *

A few days later they came to collect him. They were moving in a few weeks, from the flat to their first house, in Cumbria, on the other side of the Pennines . . . rather more suitable for a lively puppy. In the meantime, they would just have to manage.

The breeder warned them that Sam had never been in a car before and might be rather apprehensive, but she had reckoned without Sam's character. From the moment John moved off down the drive, away from the farm where Sam had been born, he was hooked. He sat, upright and rigid on his seat, staring over John's shoulder out of the windscreen, an expression of intense concentration on his face. There was no need to restrain him: he was too absorbed to think of being naughty. John said he had an uneasy feeling that at any minute Sam might remind him to stop at the red traffic light. And from that day on, it was always the same. The problem was keeping Sam out of the car, not getting him into it.

Before they reached the streets of the little town, Sam's body clock, which dictated some twenty-one hours of sleep and three of frenetic activity, had sent him into snuffling oblivion, his creamy muzzle laid across Tina's arm.

They carried him through the small courtyard and up the stairs. He slept on. Tina laid him in his new basket. He opened one eye, yawned hugely, and went back to sleep. He had come home . . .

* * *

John waved his hand out of the car window.

'Imagine this in a blizzard,' he said.

They looked out across the moor into the distance. There was a great, plunging space below them, a few houses clinging darkly to the edge. A hawk hovered, so close that they could see its stippled back. It swung out, over the cracked river, then off above the moor where the sheep seemed thrown upon the purple heather like tiny jack-stones.

'Let's get out,' said Tina. 'We've got a bit of time before the removal van catches up with us.' They had pulled into a lay-by, and let the long roar of traffic rumble past, belching diesel fumes. Now there was nothing but the sound of the wind.

'It's so peaceful,' said Tina. 'Just listen to that silence.'

John laughed.

'I know what you mean.'

26

Suddenly, from the car, there came a blood-chilling howl of desolation and the sound of paws scrabbling on the window.

'Oh, Sam. We forgot you. How could we!' Tina ran over to the car, and opened the door.

'That dog,' said John grimly, 'is impossible to forget!'

Tina was fumbling with his collar while Sam tried to follow his nose in a high speed exit out of the car.

'Oh, Sam! Do keep still. I must put you on a lead. There are sheep about.' She finally succeeded, and Sam crabbed his way over to John, greeting him as though they had been parted for years rather than minutes.

'OK, Sam, go down. I get the message. I know you've been cooped up for hours. We'll give you a bit of a walk.'

They began to walk a little way along the road, Sam pulling frantically on the lead, feeling the sharp summer wind on their faces.

'Is it always cold up here, even in the summer?' Tina shivered a little and drew her cardigan round her.

'This is the top of the Pennines,' John said. 'I think it's supposed to be the highest road in England. It's bound to be cold this high up.'

She stopped, and looked out once more over the great sweep of fell sleeping under the sun.

'I can see what you mean about being up here in a blizzard. Whatever must it be like in winter?'

John pointed ahead to a pole at the side of the road.

'You see that striped pole. When this is six feet deep in snow, that's the only way you can tell where the road is . . .'

She shivered again.

'I hope we're doing the right thing moving up here . . . it's so wild. Even Yorkshire didn't seem as wild as this!'

John laughed. 'We were living in a town, don't forget. The Yorkshire moors can be just as bleak and dangerous in winter. But the air up here . . . it must be just about as pure as you can get. When we have children, there couldn't be a better place to grow up. But nonetheless, I have to agree. There can't be many spots as desolate as this in winter. Just imagine if you were out there, lost, in that lot . . . how would anyone ever find you before you froze to death?' Tina shivered and turned up the collar of her cardigan.

'Yes, it is beautiful, perhaps because it is so very wild.' They had stopped, leaning against the gate, Sam with his paws up, nose just able to poke through the bars halfway up. He snuffled excitedly at the wind coming at them from across the fell.

John laughed and bent down to stroke the pale golden head.

'I wonder whether we'll ever make anything of you, Sam. You'll have to grow a bit, for a start. You know, I don't think he's ever going to be very big for his breed. But I reckon he'll be pretty strong. Sam . . . you can stop pulling your mistress off her feet every time she puts you on the lead, for a start!' Sam looked up at his master, eyes alert, working it out.

'Perhaps we can find somewhere nearby where they run obedience classes,' Tina said thoughtfully. 'One of us could take him. With the baby on the way, it'll probably have to be you. And he needs a firm hand. It's because he's so intelligent. At the moment he spends all his time working out new ways of getting into mischief; it would be nice to think he was using his energy more usefully! And if he doesn't stop soon, there won't be anything left in the new house he hasn't chewed up already.'

John smiled.

'Well, we can't say we weren't warned about it. The breeder made it quite clear to us what we were taking on—do you remember—the day we chose him?'

Tina nodded.

'After a few days I wondered what had hit us. And he isn't much better now.' She bent down and stroked the soft head. 'But I couldn't imagine life without him.'

28

They returned Sam to the back seat of the car and drove on over the moorland road. Ahead of them, the fells reared dark against the sky; to the right, a long sweep of bracken rose up towards the crags. Tina rolled down the window. There was that scent again: wild, and cold and free. It stirred the blood, made you feel alive.

She remembered how they had first come here, on holiday before they had Sam; how they had driven up the ancient road into the heart of the high fell. It was a country where giants had stalked, scattering scree like dust at the feet of the fells, binding the crags with mist. The names rang out of the granite like iron—Dunmail Raise, Seat Sandal, Steel Fell. Here, Dunmail, son of Owain, King of Cumbria, had fought Edmund of England a thousand years ago. The clash of sword and stone axe still echoed from the crag. They had walked through sodden bracken to the grey tarn where the crown of a dead king was said to lie. Suddenly the sun had burst out of the cloud. The grass had sparkled; granite tops had soared against a brilliant sky; the heather glittered with webs of silver. In Grisedale Tarn all the jewels of the dead king glowed for a few moments before the water turned once more to pewter, and the clouds covered the sun.

Now they were coming back to live in this wild country. Because they had both fallen in love with it. Here, where folk seemed fierce and yet were kind. Where the ancient dialect, which the Vikings had left behind, still lived on in the everyday language and in the names of the fells. Where they felt they might one day belong.

3

'What'll it be? A la'al one or a large one?' the publican asked.

'Half a Guinness and a pint, please,' John said.

Sam suddenly appeared from nowhere, planting his paws on the edge of the bar. A pink tongue snaked out, searching for a stray puddle of beer.

'Bad dog, Sam.'

'Ay, he's alreet.' The publican stretched out his hand and fondled Sam's head. 'He's nowt but a puppy.' He was a huge man, half-filling the tiny space behind the bar. His hand took in the whole of Sam's skull, feeling the bones with gigantic fingers. Sam tried to lick him.

'I know a bit about dogs. He'll make a good dog, this 'un. Given a bit of training.' He pointed to a colour photograph at the corner of the bar. It showed a brindled Border Collie, tail streaming, poised on a crag. 'That were my Lass, there. She were a wonderful bitch. I do miss her.'

'She looks beautiful,' John said. 'What happened to her?'

'She came off crag . . . in a blizzard. It were terrible. We were out looking for a fella from Grasmere. Damn fool had gone oop, by himself, in snow. That were some search, I can tell tha. She jest went ower top of crag, going full pelt. I suppose she had a scent of summat. Any road, I lost her. And we found poor bugger three hours later—in a drift. Dead. What a waste of man and dog! It allus gets you. No matter how many times you've seen it. But there's no good moaning. I've got a good new dog now. He's promising, I'll say that.'

By now Tina had joined John at the bar, despairing of ever getting a drink.

30

'Do you mean you actually use dogs on the fell? To look for people?' she asked, a trifle disbelievingly. 'I knew they did it in Switzerland. With St Bernards. But here, in Cumbria. I'd never heard of that.'

'Ay, they do. The only difference is, they don't carry brandy oop here. Reckon we're not that soft.' He nodded over to a box which stood by the photograph of the Collie. 'That's what it's all about.'

John picked it up. It was a collecting box with the initials SARDA stencilled on the side and a profile of an Alsatian.

'I haven't heard of that one,' he said wonderingly. 'Is it like the RSPCA?'

The publican laughed.

'Nowt like it. It's a bit exclusive, is this one. Hasn't been going long. But it's a good cause.'

John put the box back and felt in his pocket for some change.

'What do the initials stand for?'

'Search and Rescue Dog Association. I don't suppose folk down south have owermuch need for Search Dogs. But oop here, they're proving thesselves—working alongside Mountain Rescue Teams, tha knows. Ay, it's only been going a few years and we've a lang way to go. And there's a bit of prejudice! Some folk think it's too risky—having to rely on a dog when there are lives at stake. It's a point of view, but I think if they saw dogs working, they might change their minds.'

Tina was thoroughly intrigued. She perched herself on a vacant bar stool and John passed her drink along. A party of late walkers came heartily into the pub, smacking their hands together in anticipation of the first pint of the day, their backs bent against the low beams. It was several minutes before the publican was free to talk to them again. Sam, with a deep sigh of boredom, went to sleep at John's feet, his nose perched uncomfortably on the foot rail. All this talk did not concern him.

'Ay, as I were saying,' the publican moved along the bar towards them and began to polish the glasses. 'There aren't more than a handful of dogs in England that can be called Search Dogs. It's a tough training—takes three years at least, and that's not counting obedience. And so much depends on the character of the dog. It's a rare thing; you have to have a talented dog to start with.'

'Do you mean . . . ?' Tina leaned forward, intrigued. 'These dogs go up on the hills—fells—and look for people all by themselves?'

The publican burst out laughing.

'Nay, they're not that clever! Each dog has a handler. He stays with him and they search their own bit of fell together. The relationship is vital. It's summat like a police dog. He lives for his master, sometimes he is even part of a family, like. A family pet, you might say.'

'And are they always Alsatians?' John asked.

'Nay. Funnily enough, I were having a bit crack with one of our members at last training weekend aboot that. Using the Alsatian—German Shepherd—as a symbol seems a bit daft. We hardly use Alsatians on fell at all. More Border Collies and Labradors—although even my Lass, being a Collie, had a bit of trouble in the snow. The hair balls up and weighs them down. Labradors are best in the wet, though each handler has his favourite, of course. And you can't beat a Collie for intelligence.' He moved along to serve a customer. 'Scotland and Wales, they use Alsatians more,' he added as he worked the pump.

'So Sam . . . a dog like Sam . . .' Tina stopped, feeling foolish. 'He could train as a Search Dog, if he had the right temperament.'

The publican laughed.

'Will you have one with me? What'll it be? Same again?' They smiled and nodded. 'Nay, you can't just have any dog. Sam, now, he'd be a good starter, mebbe. And the handler's just as important. He has to be a member of a Mountain

Rescue Team. He and the dog train together, if you like. They have to know the fells. The handler has to be a la'al good climber. Otherwise we wouldn't want him. And dog has to fit with him, so to speak.'

John nodded. 'I've seen the Mountain Rescue Teams when I've been climbing, of course.'

'Does tha do a bit o' climbing then?' the publican asked, putting down their drinks on the bar. 'Now then, get this down you, and good health. Nowt like a good sup of beer to keep the rain out.'

John and he began a lively discussion about climbing in the Alps, matching mountain for mountain. Closing time came and went. No one seemed to take too much notice. Sam slept on, wuffling after dream rabbits, his paws twitching. Tina was tired. Eventually, at well past eleven, they remembered the time, and said regretful goodbyes, shaking hands with the publican.

'Come again if you're in district. We'll have a crack. I've enjoyed it. My name's Derek Atkinson, by the way.'

They introduced themselves. 'I'm John Brown and this is my wife, Tina. We moved up to Cumbria not long ago. I'm taking up a job in Social Services. And this of course is Sam.'

At the mention of his name, Sam abandoned his dreams and, brimming over with enthusiastic enquiry, planted his paws firmly on the bar once again.

'Down, Sam,' John said sharply. 'Down, boy.' With a look of reproach the dog's paws slid gently off the counter leaving a trail of beer, before he fell back to the floor with a deep sigh.

'Now that's better,' grinned Mr Atkinson. 'We'll meak a gey good dog of you yet, Sam. I'll see tha again, mebbe. And by next year I'll predict tha'll not need to be tald twice to do owt.'

*   *   *

33

It was very strange in the beginning. Folk were hard to get to know. John found his new job with Social Services challenging and demanding. They missed Yorkshire with its softer dales, and the more familiar dialect. The house would need working on for ever, it seemed. It had originally been part of an old Victorian hall, long neglected. The kitchen was huge and cold, with an ancient range in one corner. John thought about the coming baby and worried quietly. There was no garden as such, only a courtyard. Had they chosen the right place?

But there, outside the windows, were the bare fells, where the sun rose in the morning. There were great trees all round them, loud with the sound of quarrelling rooks; there was a shining road outside the gates which led up to the summit of the fell, past a field of young bullocks and a stone farm and sheep on the lower slopes. It wound away past the gates into the valley below and the village sleeping by the river. It was a good place to be born into, a good place to grow.

The baby, a boy, was born just after Christmas. They called him Matthew. Later, John remembered looking out of the window of the hospital, his son in his arms, and seeing a helicopter fly past on its way to a rescue on the fell. He had held the baby up to watch, but Matthew had only yawned and closed his eyes, contented and ready for sleep.

Sam felt left out. When they brought Matthew home, he suffered pangs of jealousy for a few days, but in the end the interesting nature of this creature, that squeaked and cried and smelt milky, was too much for him. He took to sitting by the pram, looking protective, and if the baby cried he would search out Tina, his forehead wrinkled with concern, telling her that something was wrong. It was the beginning of a long friendship. From the moment that Matthew's eyes were able to take in anything other than the faces of his parents, Sam became an alternative centre of his world. A source of warmth and comfort. Another special person to love.

34

John put it to Tina one evening. 'I'd like to start obedience classes with Sam. I haven't liked to take yet another evening, but now Matthew seems to settle down after his feed, do you think it would be a good idea? I'd certainly like to give it a try, and now they are moving one of my counselling sessions to the afternoon, it only means two nights a week out.'

The following Thursday, Sam, with a new collar and lead, needed no persuasion to get into the back seat of the Morris Minor. He knew something was up and on the journey into Penrith he sat with a pious expression on his face, like a small boy going to choir practice for the first time. John wonderd how long it would last.

Not long. Sam, scenting dogs even as they parked the car, crabbed his way along the pavement, pulling at the lead, and his entrance through the swing doors was more precipitate than John would have wished. There were dogs everywhere—Whippets perched on the stage, Yorkshire Terriers with bows in their hair being cuddled protectively by their owners, Cocker Spaniels lying moodily with their heads between their paws, Red Setters springing about wildly, pursued by lanky girls, Alsatians with thick-set men who looked as though they meant business, several Labradors, black and yellow, a Bulldog which was snoring loudly, and a huge grey Irish Wolfhound looking extremely hurt and puzzled by the whole thing.

This was the beginners' class and it was not difficult to see why. As Sam burst through the door and then stood, rigid, his tail at 90° to his body, challenging all comers, such a cacophony of barking, squeaking, scrabbling and shouting erupted that John, dragging an unwilling Sam who was by now delivering a few insults of his own, slunk off into the farthest corner by the stage, convinced that they were in imminent danger of being asked to leave. It was not until the next dog—a collie cross—had the temerity to show its face through the doors, triggering the whole performance again with Sam joining in in full cry, that John began to feel a little more comfortable. Then a small man came in followed by an extremely decrepit Alsatian without a lead. The barrage of hate began again, but not for long.

'Quiet!' the little man roared. There were a few seconds of shocked silence, owners and dogs both looking suitably chastened. Then a mongrel down the far end began an argument with the Cavalier King Charles that was cowering under the next chair. The little man, followed by his Alsatian, strode across the room.

'Keep your dog quiet, please,' he said softly. The poor owner went scarlet. Everyone else was heartily glad it was not them. John looked at Sam. He was whining softly, his ears up in a line across his head, his shining gaze fixed on the little man. John put a hand on his head—he was vibrating with eager attention. The other dogs no longer seemed to exist.

'My name is Arthur Crabtree. This is my dog Amber. Welcome to the class. I wish to make it clear that I expect you to keep your dogs under control. There will be no fighting . . .' he paused as a snarling match between two Whippets was hastily suppressed by the owners, '. . . among handlers or dogs,' (a ripple of nervous laughter) 'and no barking or unnecessary noise.' The Irish Wolfhound gave a yelp as its owner whacked it on the nose for growling at a Poodle in the corner. 'It will not be necessary for you to hit your dogs' (more red faces). 'Your hands on the lead and your voices are the only aids you will require. And later we shall use a whistle. Now, if you will get into a circle I shall come and inspect your collars and leads.'

'I say, excuse me,' a braying voice broke in. It was the owner of a beautiful Golden Retriever, which was lifting its lip at any dog that came within ten feet and which looked very nervous, unlike its owner. 'Aren't there rather a lot of us? I mean, how can we possibly discipline our dogs when there are so many distractions?' As if to prove his point, his dog dived under the chair at that juncture and a vicious fight ensued for several seconds, before the distraught owner of the Jack Russell hauled it out. The Retriever emerged looking thoughtful, with a large scratch on its nose.

'The whole point of bringing your dogs here is that you want your animal to be well behaved and utterly trust-worthy in any situation. One of his first lessons will be to learn to get on with other dogs. He will learn to obey. He will learn that anti-social behaviour will not be tolerated. And you, too,' he looked round thoughtfully, 'will hopefully learn where you are going wrong. For there are rarely bad dogs; only bad owners. Now we shall begin. Will you place your dogs on the left side of you and walk round in a clockwise direction. If your dog runs ahead or pulls behind, a firm jerk on the choke chain (correct leads for *all* dogs by next week, please) and the command "Heel" will remind your dog. Walk on now, please.'

After six weeks John could hardly believe the change in Sam. He had learnt to walk to heel, to sit on command, to 'stay' without a lead, and he was loving every minute of it. When John put on his old 'dog-walking' jacket, Sam would shoot to the front door and sit there, quivering with excitement, his eyes looking at his master as though to say 'Come on then! What are we waiting for?'

And the class as a whole! No more barking or fighting or bad behaviour, with the exception of the Golden Retriever which had been unable to settle (more the fault of the argumentative owner, John thought) and had left after three weeks. By the end of the first half-term the motley collection of dogs and owners, though still widely different, were beginning to show signs of self-control. And what a lot he had learnt about people, and about his own way of handling dogs. There were the wheedlers, the blusterers, the 'I'm not certain I ought to be asking you this, but . . .'

types, the bullies whose dogs had learnt to turn a deaf ear. John's natural affection for Sam had led him to be a bit of a wheedler. Sam, sensing this, never used to bother to obey first time and could often charm his way out of obeying at all. Mr Crabtree changed all that. John learned how to give clear, assertive commands and, strangely enough, Sam responded.

There were only two areas where difficulties still loomed. Sam, asked to 'lie down', would still lie on his back to have his tummy tickled, to the huge amusement of the rest of the class, and he still hated Alsatians! One of the exercises the class had to do every week was to weave in and out of each other with their dogs. John dreaded it. They would pass within a foot of each of the three Alsatians in the class. There would be a suppressed mumble of hate from Sam, thrown out challengingly, and a tense response from the other dog. John felt rather vulnerable in the middle. He did not fancy having a chunk taken out of his leg by an Alsatian, or any other dog. He confided his fear to Mr Crabtree, who shook his head, lined up all three Alsatians in the middle of the room and walked Sam over and over again straight through the middle of them. Sam walked to heel, looking insufferably well-behaved. There was not a whisper of aggression.

'Tension,' said Mr Crabtree, coming back to John. 'Not Sam's. Yours. You're communicating it to Sam. You're expecting him to behave badly and he does. Forget about other dogs. Think about something else. Dogs are very telepathic. Never forget that. Go on, try again.'

It was a lesson John never forgot. And Sam, given the quiet command 'Leave', would ignore even the noisiest dog and walk past to heel. Tina found she could take him out with the pram in perfect safety. Occasionally a strangled bark would burst out of him; he would look away guiltily, his ears flat with shame, and they would pass the opposition without further incident.

Some months passed. John was still taking Sam to obedience classes—by now he had moved into the Graduate Class—but there was something wrong. Sam was just not progressing. All those bright hopes John had had of him in the beginning were starting to fade. He still did as he was told . . . but without enthusiasm. Only in their regular daily practices—often a snatched ten minutes on the local playing field at the end of the day, did John sense that energy and excitement he had once felt flowing from Sam in the beginning.

But in all other respects he was still a great character. Everyone said so. They had entered him that year in the Guide Dogs for the Blind Annual Charity Show, just for fun. His legs were too short, his head too heavy and he lay on his back to have his tummy tickled when the judge approached. And it was while the judge, a rather portly man in a pin-stripe suit, struggled to look in Sam's mouth while he was lying upside down, that the Gun Dog Class finally disintegrated. Sam, still on his back, had evaded the judge's grasp with a deft flick of his head and snatched the carnation from his buttonhole. The sight of Sam, lying in the ring, on his back, tail thumping, with a red carnation in his mouth, brought the house down. Even the judge saw the joke.

Sam was something special. Everybody was his friend. He was always happy to play, to swagger around with an old slipper or his watering can in his mouth making strange grunting noises, to lick the cat's kittens, to sneak up on the tortoise and bark loudly in its non-existent ear.

He had the Labrador's strong instinct to retrieve, and now that John had confidence to let him off the lead in the woods and know that he would come when he was called, he would have a wonderful time finding sticks in the undergrowth and racing back with them, his eyes shining with pride, to drop the saliva-covered remains at John's feet.

At the end of one of the classes, Arthur Crabtree called John over and asked him to wait behind. He came straight to the point.

'I don't know what's wrong with Sam,' he said thoughtfully, stroking the dog's head. 'He started off so well. But he seems to have lost interest. Don't you feel it, too?'

John nodded. 'I know. He just seems to have gone off the boil. Lost all enthusiasm. He'll still do what I tell him, but it's so half-hearted. I keep expecting to see him yawn!'

'I think he's had enough,' the instructor said bluntly. 'I think he's reached his limits. I'm sorry. He was so promising. And I'm sure he's intelligent enough to go on. But the dog has got to see the point of it all—and I don't think, frankly, that Sam does any more.'

Sam lay down on the floor and gave a deep sigh, his eyes looking into the distance. 'See what I mean?' The instructor shook his head. 'With a dog like that—he's such a character, such an individual—you can probably only go so far. It may be that he'll never be totally reliable—there will always be that random factor. He'd never be up to trials standard, in any case.'

John swallowed hard. 'I had great ambitions for Sam.'

Mr Crabtree smiled. 'I'm sorry. It's just that if we go on, we'll undo all the good work of the last term and a half. Let's stop while we're winning. Keep up the training by yourself. Don't let him forget. He'll always be reliable enough now for the basics. Otherwise, just enjoy his character.' He looked down at Sam. Sam looked back up at him from the tops of his eyes, his forehead puckered in velvety lines.

'Good dog, Sam,' he said. 'You're a smasher.' Sam rolled on his back and thumped his tail, whining with joy. 'I've enjoyed having you.'

4

More than a year passed by. There was so much to do. Matthew and Sam were growing up together—into everything. Tina had her hands full. A new baby was born, a girl this time. Sam took up his pram-watching duties once again.

In the spring, when the intake fields were stocked with lambs, John and Tina took a day off from the endless jobs around the house—would it ever be finished?—and took Matthew, Anna (the new baby) and Sam for a drive over to Grasmere. Spring was reluctant to come this far north—whereas over in Yorkshire there would be a mist of green over all the hedges, here the buds were hardly opening. The becks were overflowing into cascading waterfalls between the crags, the whole country had an appearance of winter, and only the lambs tottering about beside their mothers, as though for Matthew's special delight, told that spring was on its way at all.

It was a few years now since a holiday in Grasmere had, indirectly, so changed their lives. Then they had had no children. Then the thought of living up here among the fells had been no more than a dream; and in those few short years they had made it a reality. Tina was still amazed that they had done it and, now they were here, she could not imagine living anywhere else. They might still be 'off-comers' to the Cumbrians, but it did not stop the natives from being some of the most warm-hearted and down-to-earth people she had ever met, once you got through that initial reserve.

41

They stopped to feed the baby in a lay-by. On either side there was a wooded gorge, with here and there a bare patch of fellside showing through. John let Sam out of the car. Ahead of them, at the base of the gorge, there was a flat area of bracken and short grass. Calling him to heel, he walked off across the broken ground, with Matthew on his shoulders. Tina watched them, smiling. This was what they had dreamt about. Sam was now so well-trained that they could go for long, long walks, back-packing the baby in her special harness.

There was a shout from John. She looked up again just in time to see Sam streaking off up the fell, pursuing two sheep.

'Heel, Sam!' He took no notice. The sheep were in a panic, bounding up the fell like kangaroos. They disappeared over a rocky promontory, John with Matthew now in his arms in hot pursuit.

Tina went cold, 'Oh, my God,' she whispered. 'Not Sam. Not Sam doing that. He'll kill them. They'll be off the crag.' She watched helplessly, trapped by the baby feeding contentedly in her arms. Visions of Sam savaging the sheep, tearing them to pieces as she had once seen on the fell, driving them off the precipitous side of the gorge; of Sam being shot without question by an angry farmer. But Sam! He wasn't like that. He could never be vicious. He could never be a sheep worrier. But that was what everybody said. She knew enough already to realise that sheep worriers were usually domestic dogs, gentle creatures by the fireside. Only sometimes, and that was even worse, were they rogue sheepdogs themselves. You could never tell.

There was no sign of John or Sam. She stared at the skyline, willing them to reappear. There was a sound beside the car. She looked out of the window on her side. There was Sam, flat out on the tarmac of the lay-by, panting fit to bust, his eyes shining, and there was blood on his muzzle. She felt sick with horror. Prising the baby away from the breast, she winded her quickly and laid her in the carrycot in the back seat.

'Sam! Bad dog!' She got out of the car. Where was John? He would know what to do. Should she punish Sam now, now he had come back? 'Bad dog! Oh, Sam, how could you?' There was blood on his whiskers and streaking down from his eye. They would have to find the farmer, pay for the sheep. If he had tasted blood, that would be the end. He would have to be destroyed. She swallowed down the tears.

John came thundering through the bracken, still with Matthew in his arms. 'I can't find that damn dog anywhere. God knows where he's got to . . .' he shouted.

'He's here. Oh, John, he's got blood on his face. What has he done? He'll have to be put down. Oh, why did you ever let him off the lead?' She began to cry.

'It's no good punishing him now. It's too late. If I could have caught up with him, that would have been the time. But, thank God, the sheep got away. They scuttled off across the crag. Sam just lost interest. But he might not next time. He was pretty intent on chasing them.'

'But the blood . . . ?' She gulped.

'Let's have a look.' He bent down. 'Sam, I'm ashamed of you.' The dog lay on his back, wagging his tail. 'It's a great game to him. I don't know how you cure it. After all that obedience training—when it comes down to it, he turned a completely deaf ear.' He stood up. 'It's all right. He must have scratched his lip on a bramble. And look, this was right up against his eye. It was lucky it didn't go in.' He held up a large thorn.

The day was spoiled for them both. Even Sam lay disconsolately in the back of the estate, his head between his paws, knowing he had broken the rules. But John privately doubted whether it would stop him doing it again. He was almost sure that the sheep were young tups, but what if they had been ewes, heavy with lamb? Chasing them would have been enough to make them abort their lambs. Ewes were easily upset, even these Herdwicks that were used to the rigours of the fells.

They drove round the lake. It was grey and wrinkled and blustered by a cold March wind. The fells were still hugged to themselves in their winter colours. It did nothing to lift their spirits. Only the sleeping baby and Matthew singing a private, made-up song, strapped in his safety seat, seemed content.

'Let's go up towards Kirkstone along those back roads where we used to walk,' John said. She nodded, her mind still preoccupied. 'We could find a pub where they don't mind children and stop and have lunch. OK?' She tried to smile. They would never be able to let Sam off now. He could never be allowed to be free on the fells. Always, in the back of their minds, would be the knowledge that he had once chased sheep and might do it again. She saw again that intent expression on his face, the way he had bounded up the fellside at the heels of those terrified sheep. 'Off-comers.' 'Townies with their dog.' 'Don't know how to behave in the country.' She could hear it now.

She looked out of the window. John was slowing down, pulling into a pub car park. There was something familiar about it.

'Thought we might try here.' John was looking inscrutable. He pulled the car into the space. 'It's the Carpenter's Arms, where we went that night, just after we arrived. I wonder if Derek Atkinson's still here.'

'John, did you plan this . . . ?' She looked at him.

'I phoned up this morning and booked a table as a surprise. But a woman answered. It may have changed hands.' He smiled.

'He won't remember us anyway.' She lifted Matthew out of his seat and Tom slid the carrycot out of the car. They walked through the door marked 'Lounge Bar'.

'Is it all right . . . for the children?'

John nodded. 'Yes, I checked. You know what pubs are like in Cumbria. Very easy-going. Go on in.'

There was a bright fire in the grate. Red covers on the battered old settee in the corner. Otherwise it was just as Tina remembered it. And a familiar face behind the bar—Derek Atkinson, beaming at them.

'Now, I thowt it could be anyone, and there are that many Browns about. I jest wondered when the wife told me about the reservation. But what's this? Two la'al bairns already? I can't believe it! Ay, it's nice to see you.'

He shook them both by the hand. 'Ay, coom on through to the little snug. I'll get the bairn some lemonade and you'll have one on me. This is an occasion.' He reappeared a few minutes later with a tray of drinks. 'We'll have a bit crack before it gets busy. I had a feeling that we hadn't seen the last of you both, and I were spot on! Hey, where's Sam? Tha can see, I hav'na forgotten. Has owt happened to him?'

John shook his head and laughed.

'No, but we thought if people were eating, we oughtn't to bring him in.'

'You bring him through here. I want to see him,' Derek Atkinson grinned. 'Is he still as daft? Or have you done owt with him?'

John went out to the car and fetched Sam in. He walked quietly to heel and when told, 'Go down, Sam,' he dropped immediately to the floor of the bar by John's chair.

'Well, well, so you've been working on him. I can see that. And he's grown into a fine dog.' Sam was meanwhile spoiling himself by squirming along the floor towards Derek Atkinson's chair. He looked down and laughed. 'But he's not perfect yet!'

John made a rueful face.

'It's been a problem, Derek. I took Sam to obedience classes over at Keswick. He did all right to begin with . . . got the basics. Then I don't know what happened. He just seemed to lose interest. I don't think the instructor was too impressed.' Sam gave a deep sigh. Derek shook his head and smiled.

'Well, if you have a spirited dog—with a bit of intelligence, you know, likes to do things his way—he can either take to obedience like a duck to water, or it can crush the spirit out of him. Take the SARDA dogs. The official policy is that potential dogs shouldn't go for formal obedience training. Sometimes it stops them thinking for themselves. They look at it this way: a dog you can rely on to take the initiative in a white-out shouldn't be too tied up with formal training. Mind you, SARDA require a good standard themselves, and some dogs certainly have started off with going to classes. But it sounds like you might have a Search Dog in the making there . . . wanting to do his own thing, as they say.'

Sam snored loudly, wuffling his cheeks in a dream. Derek shook his head.

'What I want to know is, why is it every time I see your dog, he falls asleep at me feet!'

While they supped their drinks he told them that his own young Collie, which he had been beginning to train when they saw him last, had just been graded as a Search Dog. 'Ay, we're very proud of Lad. I think in time he'll be as good as the bitch we lost on the fell.'

'I bet he doesn't chase sheep,' said John ruefully, and proceeded to tell Derek Atkinson all about the morning's episode.

'Now we're saddled with a dog we can never let off,' he said, shaking his head. Sam looked up at his master with liquid, penitent eyes, and put his head down between his paws. 'And yet look at him. He would never harm a living thing. It's sheer mischief. They run away, so he runs after them. Now he's started, I can never let him run free. Once he was off, he just ignored me.'

Derek Atkinson shook his head. 'I know. I know. It's a

real problem in the fells is sheep-chasing. You're got to have a trustworthy dog that won't look at stock under any circumstances.'

'If only there was a way of curing him,' John said. 'It's no good punishing him after he's come back—or next time he won't come back. I don't know—what he needs is some kind of aversion therapy.'

Derek Atkinson looked thoughtful. 'I tell you what: I might be able to help. I'll have a word with a friend of mine. Give me your telephone number and I'll phone you in a few days. It's a shame to see a fine dog like Sam forever on a lead because you can't trust him on the fell. It's a life sentence. I'll see what I can do. Now, I'd better go and help Ivy serve the lunches. I'll see you before you go.'

'Fancy him remembering us after all this time—and Sam!' Tina said after Derek had gone through to the other bar.

John grinned. 'That's Cumbrians for you. Long memories, slow to anger, slow to friendship. Steady as a rock.'

Derek Atkinson's wife came through with plates of hot steak and kidney pie.

'I've made a special helping for the bairn,' she smiled. 'I'm Ivy, by the way. It's nice to meet you. How are you enjoying living in Cumbria? It's a wonderful county. But the folk do take some knowing.'

Tina smiled back.

'Thank you. Everyone's been so kind. We love it. We don't even mind the rain—too much.'

When they had finished, Ivy came through again to collect the plates. Baby Anna was asleep in her carrycot on the bench. Matthew was on the floor, playing with his lorry. Tina and John, too full to move, were staring into the fire.

'There! Now is that better? It's a thin wind out there today. I wouldn't be surprised if we don't get more snow, even though it is supposed to be spring. Derek's just closing

47

up but he says you're not to move yourselves. He's bringing through some coffee.'

Tina looked at her watch.

'Goodness. Is it that late? Where on earth has the time gone? We must have been dreaming. These last months have been so hectic we've hardly stopped. It's nice just to sit quietly for a little while.'

Ivy nodded sympathetically. 'Grab your chances while you can. They're rare enough with a young baby. I know. I've reared three sons and each one was a handful.'

Derek came through from the bar with a tray of coffee cups.

'Now,' he said when he had settled himself. 'There was something I wanted to talk to you about, John. Have you thought about becoming a member of a Mountain Rescue Team? They're always short of experienced climbers, and I remember you telling me you'd spent quite a bit of time in the mountains.'

'I had thought about it,' John admitted. 'But I didn't want to go crashing in. The Mountain Rescue Teams seem such a close-knit bunch—it's very Cumbrian, if you know what I mean . . .'

'Ay, you've got a point. I know exactly what you mean,' Derek nodded, looking into the fire. 'I'm lucky. I was born and bred in fells. You get accepted, like, from the beginning. Everyone knew my father. Everyone knows me. It's as simple as that. Tha doesn't have to prove thaself.'

'Anyway, I didn't want to rush in,' John finished. 'We're very new.'

'Ay, you'll be off-comers for about thirty years if you're lucky,' Derek laughed. 'They won't like you any the less but they won't forget! Matthew, now, it'll be different for him. And things are changing—new folk are moving into the fells all the time. Folk here will have to change with the times.'

'I don't specially want it to be any different,' John said

thoughtfully. 'You feel the same, don't you, Tina? It's like another country. Not closed in, shutting out strangers. They couldn't be kinder. But separate. It would be stupid to think you could be a Cumbrian after a year, two years. It's a slow place, to grow with.'

'It must be the rocks,' Tina laughed. 'I feel the same. That's one of the things we love.'

Derek was listening attentively. He nodded.

'Well, I've been all over the world. In the Navy. But there were never any question—this is where my roots are. Simple as that.'

Ivy smiled. 'More like a stubborn old tooth you can't pull out!'

'Anyway, I think it would be a good thing, if you are interested in joining a team, if I had a word with Bill Nicholson over at Langenhow. I'm sure they'd be more than glad of your skills—if you match up, mind. They have a training night on a Tuesday. Just be prepared, if you go down. You know what the sense of humour is around these parts. Well, it's worse over at Langenhow. They're downright cruel!' He laughed. 'But they do sup a good pint.'

Later, when they were driving over Kirkstone in the wind-torn evening light, John shook his head.

'Why do I just get the feeling I've been invited to join some exclusive London club?'

She laughed. 'Cumbrian club, you mean. And I bet the initiation ceremony is a lot tougher!'

Tina was right. A few days later Bill Nicholson telephoned John. Would he like to come down to the Fox and Hounds in Langenhow at seven o'clock on Tuesday? They could have a pint and a crack just to see how he matched up.

'Charming,' said John when he recited the conversation to Tina while she was cooking supper. 'Nothing like making you feel wanted.'

'Just think about it, John.' She waved the fish slice at him. 'They must have any number of fools trying to get into

49

Mountain Rescue Teams, seeing themselves as shining heroes up on the crags, when all they've ever done is to walk up Scafell Pike a couple of times. They can't afford to make mistakes.'

John drove down to Langenhow the following Tuesday evening. It was a prematurely dark, miserable evening, raining steadily. He had prudently placed his boots and cagoule in the back of the car.

The lounge was deserted. He walked up to the bar. A man was polishing glasses.

'Excuse me. Where would I find the Mountain Rescue Team?' The man looked up and winked, then nodded laconically towards a little door labelled 'Private', on the left hand side of the fireplace. He went through. It was a small room, full of smoke, although through the haze he could only see two wizened old characters sitting by the blazing fire at the far end, one of whom was puffing violently on a pipe while the other had a cigarette hanging dangerously from his lower lip. They were both clutching pint pots. The one on the left looked up from his contemplation of his pint.

'I'm sorry. I was looking for the Mountain Rescue Team.' John made to go out of the door.

'Is tha John Brown, then?' The figure on the left stood up. He was somewhat eccentrically dressed in an old shooting jacket, breeches and leggings. 'I'm Bill Nicholson. Was tha expecting summat else? A young strip, mebbe?' He chuckled. 'I dessay, I'm past it now. Wouldn't tha say, George?' The other seemingly ancient occupant of the room got to his feet.

'You're not strang, Bill. I have to say that. It's time tha was put oot to grass.'

John stood there, feeling uncomfortably that the joke was somehow being laboured at his expense.

'Come in and sit thisself down,' Bill Nicholson said kindly, obviously feeling that honour had been satisfied. 'What'll tha have? A pint? We've time for a crack before young fellas coom in and spoil the quiet.'

For the next half hour in the stuffy little room, John found himself subjected to a casual sort of third degree. So casual, indeed, that it was not until he got home that he realised how much Bill Nicholson had squeezed out of him with his quiet, almost absent-minded questions. As he probed intently into the dark recesses of his pipe with a match, or supped his beer and stared into the fire, George, too, would take the opportunity to interject a comment or a question. The conversation veered in an unstructured sort of way round to mountains. John began to describe a medium-difficult climb he had made in the Alps the year before. They nodded sagely, a wondering look on their faces. John was beginning to get a feeling about these two. He had a suspicion they had done the lot. They just weren't telling.

The other members of the team began to arrive, one by one. The room filled up and became more smoky. John was introduced around. One young man, Geoff Hornsby, was introduced as the team leader. John was surprised.

'I thought Bill was the leader.' he looked round at Bill, who was standing at his elbow with the remains of his pint.

'Ay, it's sad, is that. Poor owd Bill can't keep up with us young folk the way he used to. He's a bit of a father figure, you might say.' Bill growled good-humouredly and moved away. Geoff Hornsby looked at John and gave a huge wink.

'Don't you believe it. If you ever have the misfortune to be paired with Bill, you'll be flat out by the morning and he'll be springing about like a spring lamb. He and George Parkinson now, they climbed every peak in the Alps and several in the Himalayan range before they were fifty. It were Bill and George started the Mountain Rescue Team way back—informal, like. We couldn't do wi'out them. They know every crag and corrie of the fell for miles around. Don't be taken in by old Bill. He'll pull your leg unmercifully. He's that sharp, I tell him, he'll cut his own leg off one day. But he's a gey good mate to have up on fell, when t'wedder clamps down.'

'Do you think I'll be OK?' John asked anxiously.

'Tell tha what. Coom along now. We're off to HQ for a lecture on first aid and a bit of practice. We can talk about it then.'

* * *

'How did you get on?' Tina was sitting up in bed waiting for him.

He grinned. 'I think I'm in, if only for a probationary period. That was a social get-together—it only happens once a quarter. Afterwards we went back to their HQ—you should see it! It's a converted barn where they've got a radio control, and all the equipment. Good bunch of blokes, too. I'll be lucky to get a place on the team. It must be a popular one to join, and they're relatively prosperous, being on the edge of the lakes. I reckon it's still a struggle to raise funds, though.'

'Did they mind—your not coming from Cumbria, I mean?' she asked.

'Well, they sussed me out as a stranger straight off. But you know what it's like; it won't make any difference to the way they treat me.' He paused, reflectively. 'I don't know. Perhaps I'll get my leg pulled a bit more. But I shall just have to be on my mettle. Not make too many mistakes.'

'Is it every Tuesday, then?'

'No, once a fortnight. Usually they meet at the base of one of the fells. Have a practice. Doesn't matter too much about the weather, Bill said. The worse the better. Then we repair to the pub to thaw out!'

'Sounds great fun,' she said wryly.

'Well, it's doing a good job. Apparently the Mountain Rescue Teams don't get any money from central Government. Some from the local authority, but most they have to get by flag days and fund raising. I must say I've never thought about it when I've been climbing. I've been fortunate enough never to need rescuing, but if I did I'd be very glad the teams were on call. Bill said that some of them have a real struggle to get their equipment. A lot of it is pretty expensive and sophisticated—radio, collapsible stretchers, Land-Rovers. It's high-powered stuff now, you know, even if in the end . . .'

'It comes down to a few men alone on the fells, searching in the dark,' she finished for him. He nodded.

'Or one man and his dog,' he laughed. 'Hey, guess what? There was this chap, Bob Irons, he's just moving across to Marrowdale. He had his dog with him—a beautiful Border Collie bitch called Bess. She's training to be a Search Dog. At the moment she's at the Novice grade. That will mean she can go out on searches but she generally has a fully trained Search Dog with her. There is an annual grading course, apparently, run by SARDA, for all the dogs currently under training. Bob is hoping Bess will get graded this year as a Search Dog. He's put a lot of work into her.'

'Is that what the dogs are called when they can search properly?' she asked.

'Yes. It's a pretty tough course but it sounds really interesting and challenging. For the handler and the dog. And what a fantastic end-product! A dog you can really trust on an actual search. That would be something.'

5

Before John had a chance to go to the first proper team practice, he had a phone call.

'This is Alan Armstrong over at Seathwaite. Derek told me tha had a la'al problem with tha dog. Is that so?'

John, who had been in the middle of beans on toast before rushing off to his evening clinic, took a few seconds to adjust.

'A problem? With Sam? Oh, yes, do you mean with his chasing sheep?'

'Ay, that's it. And tha wants summat to teach him a lesson?'

'Yes. Well. I want him to stop.'

'Aye, they're a menace on fell—dogs that chase sheep, tha knows. Yows abortin', brekkin' legs, that sorta thing.'

'Yes, Mr Armstrong. I know. That's why I want to do something about it.'

'Well, I can help tha, mebbe. If tha bring dog over fell to my farm, Greystones—Saturday after dinner, about two o'clock. Will that be alreet?'

'Thank you, Mr Armstrong. But what . . . ?'

'I'll see tha then. And can tha bring out a long bit of rope? For a lead, like?'

'Yes, I will. I certainly do want to stop Sam . . . but what exactly . . . ?'

John went back to his cold baked beans.

'Who on earth was that?' Tina asked.

John shook his head. 'A farmer called Armstrong. He's a

54

friend of Derek's. Offered to sort Sam out—you know, the business of the sheep chasing. It's very kind . . .'

'You sound worried.'

'I just hope it isn't too drastic. I've got a feeling Sam's in for a nasty surprise.'

*   *   *

'Greystones Farm? You're nearly agin it.' John had been driving round Seathwaite and up and down the lanes for almost an hour, cursing himself for not asking directions from Mr Armstrong. He drove down the track indicated by the helpful passerby. There was a small sign nailed somewhat precariously to the gate. 'Greystones Farm. Leave mail in box.' It was only as he passed within a yard of it that John was able to read it at all, encrusted as it was with nameless substances and overhung by tendrils of bindweed which was just producing a new crop of leaves.

'Really! How do these people expect anyone to find their way around?' he thought to himself. But he already knew the answer: everyone would know Armstrong of Greystones—everyone for miles. It was that exclusive club again, and one of the reasons he had been too proud to ask the way at the Post Office, and only now in desperation had enquired at all. It was that look they gave you, the one they reserved for strangers. Not unfriendly, mind. Just different.

It was half past two. It had been raining all morning. The track was a couple of lines of tyre marks in a sea of mire. The mixture of slurry and cinders had liberally peppered the grass verge. An ancient and unidentifiable piece of farm machinery lay abandoned by the side of the track. The backs of the barns had that dilapidated air of very old buildings which had been cobbled together again and again, and had hardly the strength to hold themselves up. John guessed, however, that they would still be there in a hundred years, doing their martyred act. It was all to do with the contradictions of this little country. Like the battered Land-Rover standing in the yard as he swung round the corner, and the brand new pale blue Range Rover tucked away, shamefaced, in the hay barn.

'For the wife,' a voice said, as John got out of his car. 'I'll tell you, I've taken a lot of stick for that car. Pooder blue, pooder blue, that's a gey good colour for a farmer! It'll be pooder blue wellies next! I hardly dare show me face in the Hare and Hounds.'

John grinned. Alan Armstrong was a tiny man, vibrating with energy. He was dressed in a pair of encrusted green cords (John was reminded of the gatepost) and a short-sleeved shirt, although there was a bitter northerly wind hooking round the corner of the yard.

'I'm sorry I'm so late. I had a job finding you.'

'I thought, as soon as I put 'phone down, I should ha given you directions. But the wife said, he only has to ask. All the folks round here know Armstrong of Greystones. There have been Armstrongs here for five generations.'

There was a scratching from the rear of the Estate. Sam, his nose pressed passionately to the back window, was sending out frantic messages.

'May I let Sam out?'

'Ay, do that. But keep him on lead, will you? There are sheep all over road and he hasn't learnt his lesson yet.'

John lifted up the rear door, told Sam to sit and slipped the choke chain over his neck.

'He knows his manners, then,' said Alan approvingly. 'He's a nice dog. I hear you might be thinking of training him as a Search Dog. Is that right?'

'I hadn't actually said anything to anyone. In fact, it seemed a bit presumptuous to even think about it.'

Alan laughed. 'Well, I can tell tha Derek Atkinson has a bee in his bonnet about Sam. He wouldn't be going to all this trouble, otherwise. Mind, he's a man who knows his dogs. And I know he thinks rather highly of Sam. Thinks he's got potential. Funny, I'll swear he said you were thinking of training Sam for rescue work. Where could he have got that idea from?'

John grinned. 'He must be telepathic. I have to admit to day-dreaming about it, but you know how it is . . .'

'Ay, you need someone to tek an interest. Especially when you're new. Folk round here can be a bit distant with off-comers. But Derek now, he'd give you all the encouragement you'd need. I should have a go, if I were you. What can you lose, anyway? But coom on, let's get this over and done with first.'

They walked together up a small lane at the side of one of the barns. There were sheep grazing malcontentedly in the sodden stubble in the intake fields.

'Ay, we've had a rotten winter,' Alan grunted. 'I have an idea aboot Sam. If we tek him through into the field with the yows—you'll want to have him on a long lead. Now some folk believe in shock treatment for dog. Put him in wi' a tup. Get him knocked aboot a bit.'

John swallowed. 'Is that what we're going to do with Sam?'

Alan stopped and ran a hand over his forehead, looking down reflectively at Sam, who was standing to heel, looking as good as gold.

'Trouble is, sometimes that treatment'll work, sometimes not. Sometimes dog'll get a grudge against sheep, like, afterwards, and he'll never lose it. And sometimes, I've seen it meself wi' a sheep dog—he's that frightened of sheep ever after he won't go near the devils. Ruins him as a working dog. It's a terrible thing to see. And you want to train that dog of yours to work on fell . . . mebbe?'

John nodded, stroking Sam's head, feeling the vibrations of excitement as he scented the sheep beyond the gate.

'Well, how would it be if he won't go near a sheep because he's that scared?'

'It would be terrible,' John said. 'But is there any other

way? It's got to teach him a lesson for good. Otherwise I might just as well give up now.'

'Ay, I know that,' Alan smiled. 'I thought we'd try this. Have you got long piece of rope I asked you to bring? Would you like to tie it on Sam's collar and then we'll go through gate, in with the old yows. It may be he'll just learn to ignore them if we spend a bit of time walking in field. It's only curiosity in the beginning. Anyway, we'll see . . .'

John, with Sam attached to the length of rope as well as his lead, followed Alan through the gate. The field was stocked with ewes and their lambs.

'If he runs after them . . . jerk him hard. Pull him on to his back. That'll teach him.'

'Heel, Sam,' John said. They began to walk through the sheep, who hedged off nervously. Sam was quivering all over with excitement, dying to chase.

'Let him run free,' Alan said quietly. John took off Sam's ordinary lead but left the rope attached to his collar. With any luck he would not notice it. Sam, feeling the freedom, shot off like a bullet, making for the sheep. Alan lifted his hand to give the signal for John to jerk him off his feet, but Nature, in the form of an elderly, almost toothless ewe, took over. She saw Sam, who was just approaching the flock in a series of frantic leaps and bounds, as an obvious threat to her lamb. She put her head down, beat a frenzied tattoo with her front feet, and then, with a strangled bleat of rage, charged straight at him. The full force of her bony strength caught him on the flank. Sam, with a howl of terror and puzzlement, was propelled several feet into the air before landing on his back in the corner of the field, hard against the wall. The rope was jerked out of John's horrified hand. The ewe gave a snort, shook her ancient coat at her enemy and galloped back to her lamb, which was chewing on a stubble patch, unaware of all the drama. Sam lay on the ground. Alan placed a restraining hand on John's arm.

'I think Nature and that owd yow did our work for us,' he said quietly. 'They can be reet fierce, the owd yows, when they've hoggets to protect, no matter if the hoggets are big enough ter fend for theirselves.' He chuckled. 'Sam's nowt but winded. She may have a gey good head for butting but not horns. She'll not damage him.' Sam, meanwhile, looking a little dazed, was getting to his feet. 'Call him to heel now, John. Then we'll walk reet through the yows.'

Sam staggered over to John, stopping every few yards to nose at his side in a puzzled way. He cast a couple of sidelong glances at the sheep, but he was going to keep his distance: they were altogether too unpredictable. John ran a hand along his flanks to make sure there were no cracked ribs. Sam was rather muddy from landing on his back in a sludgy field but otherwise there did not seem to be any damage, except to his injured pride.

'Nay, he'll not chase sheep again after that experience,' said Alan drily. 'But we must make sure that he's not frightened of them either. You can take off that rope. We won't be needing it any more. Just keep him to heel, John. We'll show him how to walk through sheep and behave hisself.' They walked quietly through the flock, which, by some mutual telepathic sense, seemed to know that the danger from Sam was over. Sam walked, a little subdued, at John's side, his eyes looking everywhere but at the ewes. But by the fourth time, he had begun to recover his old bounce, and walked along with his tail wagging.

Later, in the kitchen of the farm, drinking great mugs of tea and eating scones home-made by Alan's rather gorgeous wife, Sam lying down by his feet, John said:

'I'm very grateful, Alan. I've got a feeling I won't have any more trouble from Sam. And after we'd walked through the ewes a few times I don't think he's going to be worried by them, either. It's the strangest thing I ever saw.'

Alan laughed. 'Well, he had a harder lesson than we had planned for him but it was no bad thing if it's turned out right. I wouldn't expect to hear he'd ever chase a sheep again. More likely he'll pretend they don't exist—and you'll find the feeling will be mutual. How a yow knows which dogs to fear and which not, I'll never know. A yow'll move for one of my dogs when he asks her, but another time when Shep or Jet are running free, she won't taak a blind bit of notice. And it'll be the same with Sam, you mind.'

John got up to go, smiling his thanks at Alan's wife.

'Well, I don't know how to thank you. You've been very kind. And I've got a feeling that bone-headed old yow has done her bit as well. I've no doubt I'll have occasion to be grateful to her, too, in the future. Although Sam might not echo my sentiments just at the moment!' Sam woke up, stretched himself, found it hurt, licked himself, and rolled over cautiously.

John bent down and fondled his ears. 'Oh, Sam. Trust you to get the most bad-tempered ewe in Lakeland. It was just your luck, wasn't it?' Sam opened one eye, shut it again, and went back to sleep.

6

John began his probationary period with the Langenhow Mountain Rescue Team. Every fortnight they met at a pre-arranged spot on the fells for search and rescue drill. They practised map-reading by torchlight in the rain and the dark, while climbing a slope or, more trickily, coming down. They took turns to be manhandled down the fellside on stretchers. They talked to each other on the short-wave radios . . . not always strictly according to the book.

Occasionally, there would be a lecture in the 'barn'—on avalanches, or the use of helicopters in rescue operations, or snow structure, or any one of a hundred technical subjects they might need.

It was a marvellous antidote to work, which was often exhausting. Spring gave way to early summer. There was still a nip in the air. One evening in June, when they were halfway through supper, the phone rang. It was a call-out. A young climber lying injured. Rendezvous at Garth How.

Tina ran out to him with a flask as he reversed the car round in the road. It was raining, lightly. A chilly wind. He put his foot on the accelerator, hoping he would not meet Stephen Hodgkin's prize Friesians round Gateling Corner, as he had once before. The road was slippery with signs of their passing, but the road, thankfully, was clear. He shot through the village and over the bridge. The light was going fast. He turned left down the narrow track which led to the barn, pulling smartly into a gateway to avoid the team Land-Rover which was rattling towards him. Bob was in the driving seat. He slid the cab window aside.

'It's all right, mate. You haven't missed the boat. We're the crash party. See you on the fell.' He shot off. John caught a glimpse of Bess in the back, her nose pressed questioningly to the window.

The rest of the team were gathered in the barn, putting the finishing touches to the second Land-Rover, which carried most of the equipment. The 'crash' party (or advance party) would only have radios and basic first aid equipment. It would be their job to locate the casualty as quickly as possible and to radio back what equipment would be needed.

The others piled into the second Land-Rover. Geoff explained that he would brief them as they went along. If the casualty was where he should be, it should take the crash party about an hour to get to the spot. It was important for them to be there, at the base of the fell, with equipment to cover all eventualities, ready for a quick follow-up.

They bucketed down the track. Geoff bellowed back to them the basic facts as he had had them from the police some half an hour before. It was the appearance at the local Youth Hostel of a young walker, alone, exhausted and with a badly ricked ankle, which had led to the police being alerted. The young man said he had left his companion on the high fell. They had been walking close to Doon Crag when his friend had slipped and fallen off the crag, ending up against the base of the rock. The other man had climbed down to him, to find him unconscious and lying on his back, oddly twisted. He had scrambled back up and clambered down the northern path where he had slipped and turned his ankle. The police had called in the Langenhow team immediately.

Although this looked like a straightforward rescue, with a good map reference for the location of the casualty, they had to be prepared for anything. Bess, in the advance party with Bob Irons, would ensure an increased search capacity.

The rain was falling steadily now, and it was cold. There would certainly be a risk of exposure . . . perhaps even hypothermia.

They drove fast up to the top of the narrow pass which linked Garth How with the valley on the far side of Langenhow Fell. The tourist season had only just begun, and the road was almost empty of cars. At the top of the pass, Geoff pulled off the road and switched on the radio.

'Langhow Control from Langhow One, for a radio check. How do you read? Over.'

There was a pause. Geoff slid the window to one side and stuck his head out into the darkness.

'Can't see a damn thing.' He repeated the message. There was a crackle.

'Langhow One from Langhow Control. Reading you strength five. Over.'

'About time, too,' Geoff muttered.

'Langhow One from Langhow Control. Request situation check. Over.'

'Langhow Control from Langhow One. Bob and Tony have just started up the fell. Over.'

Geoff, satisfied, reported that they would be with Langhow Two at the end of the track in ten minutes. They swooped down into the darkness, leaving behind the last smear of westerly light on the horizon, and also losing radio contact once more as they skirted the base of Tarn End. A splatter of rain hurled itself at the windscreen from where it had been skulking behind the crag. The full force of the wind hit them, and Geoff hastily slammed the window shut.

'Fine night for June, lads,' he shouted back at them. John thought of the man lying injured on the fell—as the others were probably thinking, too — while the Land-Rover lights picked out the hunched shapes of sodden sheep beside the road. He won't even know yet, thought John, that anyone's looking for him. He won't even know if his friend has got down safely. He may be too far gone to know anything.

They found the track and roared along it in the dark, sending up a wave of sludgy mud on either side. The other Land-Rover was parked in the bracken at the far end. Jim Hodgkin, who ran the shop in Garth How and was brother to Stephen Hodgkin at the farm, was sitting in the cab, trying to contact the crash party.

'Any luck, Jim?' Geoff slammed open the back of their Land-Rover, the wind catching at it when it was half-open.

'No good,' Jim bellowed out of his window. 'They must be under the shadow of the gully about now. It'll be no good till they get out on the far side.'

'Well, we can establish contact from Control now. I want you to go up to the top of the pass with Langhow Two and act as a relay with base. It'll be no good getting directly through from here with Langenhow Fell in the way. Get up there a bit sharpish, will you? As soon as Bob Irons comes through I shall want base to arrange a chopper on standby at RAF Boulmer.'

'OK, Skip.' Jim started up his engine. 'I'll come through to you as soon as I get to the top of the pass.' He shot off into the dark, spraying Bill Nicholson with fine mud as he reversed onto the track.

'Here! Mind me bacon sandwich. That were me tea,' a furious shout went up from Bill. 'How am I going to keep body and soul together for the next six hours, I'd like to know?'

They began to unload the equipment, checking everything as they went. They would be bound to need the stretcher, which would be carried up the fell in several parts. John fastened his rucksack to a section.

Geoff leaned over the set.

'. . . Reading you strength five . . .' He listened for a few moments, the map on his knee.

'OK, everyone.' He climbed out of the cab with the map,

and spread it on the bonnet, shining his torch. They gathered round.

'Bob's located him, here.' Geoff drew a circle on the map. 'He's in a fairly bad way—wet, cold, suffering from exposure. And there's definitely a back injury. He can't move his legs. Tony Moore, who's up there with Bob, is pretty good—he's a doctor down at the cottage hospital. So he knows what he's talking about. We'd better have that special back restraint so we can strap the casualty into it and he can't shift about at all.'

Bill Nicholson leaned over the map and gestured with the remains of his sandwich.

'If we've got to lift him off, Skip, how are we going to do it oop at Doon Crag? You know what it's like oop there. There's that overhang. You'll nivver get chopper to go in.'

Geoff nodded. 'I know, Bill. You're absolutely right. I can't see we've any choice but to carry him down to . . . here.' He drew another circle. They strained to see. 'That promontory on top of Shepherd's Ghyll. It's as flat as we'll get it up there. It means carrying the casualty about a hundred and fifty yards. We'll just have to be very careful.'

'But there isn't even a whisper of a path up to Doon Crag. Only sheep tracks,' someone else commented.

'Mostly scree and bracken,' put in Bill. 'It'll be lethal in this wet.'

'We .. e .. ll,' Geoff hesitated. 'There's the long path which leads over from Tarn End. But that means a trek round. We would come out above Doon Crag, here. But it'll add another hour to the operation, both ways. And we'd have to find another spot for the chopper.'

'Best get him down quickly, poor devil,' Bill grunted. 'The longer he's oop there, the wuss it'll be fer him.'

There was a silent consensus. They checked the equipment again. Geoff decided they should take the Reviva— the hot air machine—just in case of hypothermia. The rain was coming down in sheets. It was amazingly cold. They set off up the track, leaving Geoff behind to liaise with Base Control, Bill leading them through the bracken. While they were on the flat it was fine, the well-used track following the contours of the valley floor. But then they began climbing towards Doon Crag which hung above them somewhere in the darkness, and after a few yards, suddenly, there was nothing.

Gradually it grew steeper. John was glad of all those exercises which had so strengthened his thigh muscles, and accustomed him to climbing almost blind in the dark, in that most lethal of all terrain, wet scree masked with bracken. Twice he turned his ankle and, from the muffled exclamations in front of him and behind, he was not the only one.

They climbed steadily, using their individual torches and helmet lights, nominally in pairs, but in reality straggling in a human chain up the fellside, each man protected in front and at the rear in case of falls. Their own small circle of light lit up the tightly furled young bracken which was only now breaking through the old red stuff on the heights, and shone on the half-hidden granite boulders underneath, slimy in the rain. Apart from that, the faint glow of the torches in front, muffled breathing, a few curses, and the flicker of Bill's more powerful light as it caught the bluffs and curves of the fell somewhere ahead. And always the wind hurling slaps of rain onto their backs.

Suddenly, John was aware that the chain had stopped. He stood still and listened. He heard, somewhere out to the right of them, the distant barking of a dog . . . and almost simultaneously, as Bill shouted, he saw the distant beam of a torch shining up into the rain.

Moving crabwise across the broken fell, they made their way towards the light. It disappeared for a few seconds. Then it reappeared, much closer. They clambered over a hump in the fell. There was the 'crash' party—two figures illuminated by the powerful lamp, crouching over a bright orange casualty bag, the flame of a primus stove flaring up. Bess bounded down the fell to meet them, greeting each of them in turn. Bob stood up, squinting against the light. The other man, Tony Moore, was checking the casualty's pulse, and for a few seconds did not move. Then he too stood up, wincing at the stiffness in his legs.

'You were quick,' he said to Bill as they gathered round, already busy unloading the stretcher.

'It's all the courting I used to do oop here,' Bill chuckled. 'I knows this fell like the back of me hand.' He was busy fitting together the parts of the hot air machine by the light of a torch.

'Do you reckon tha'll need this contraption?' he said, as Tony squatted down next to him. 'Can tha spare a moment?' Tony nodded without speaking, and together they made their way over to the far side of the light. There was a quick discussion. It was basic policy never to discuss the state of a casualty in front of him . . . or even the condition of his friends. Morale was an important factor.

The rest of the team, with the exception of Bob, who had taken the place of Tony by the casualty's side, came over after a couple of minutes.

'The lad's pretty bad,' Bill said in a low voice. 'Shock and exposure. An irregular pulse. Back injuries suspected. No obvious broken bones except for a badly sprained wrist. Tony's dealt with that. Ay, I think a quick whiff of the Entonox wouldn't do no harm. He's conscious on and off.' There was a groan from the casualty as though to echo Bill's words. 'The key word is *shock*. Let's just keep him calm and in a stable condition. We'll get him into the back restraint and then on the stretcher. John, will tha stay by his head and just have a bit crack with him, like? Getting down to Shepherd's Ghyll is going to be a la'al bit tricky.'

'You can say that again,' someone muttered.

'OK, OK. Let's get the lad down as fast as possible. And that means SLOW!'

It was a hazardous journey down—dangerous and slippery at the best of times but now, with the dead weight of an injured man, to whom every jolt was traumatic and painful, infinitely more difficult. For the first section John walked by the casualty, trying to reassure him. He had regained consciousness and was fretful.

'OK, mate,' John said. 'You're going to be all right. We're getting a helicopter to pick you up. You'll be in hospital soon. They'll fix you up.'

The man struggled to move against the back restraint. John laid his hand on his arm.

'I should lie still if I were you. No sense in making yourself uncomfortable. We're taking care of you.'

'My name's Ken. Ken Cullom. I fell down somewhere.'

'Yes, we know. It's all right now. You don't have to worry any more. We're the Mountain Rescue Team who came to find you.'

'I'm not hurt. Only my wrist. I'd be quite able to walk.' He had a soft accent—Irish, probably. His voice was very weak.

'You'll be up and about soon enough. Just take it easy while you can.'

There was a silence. John, trying to negotiate a slippery patch, shone his torch obliquely and rather unsteadily so as to show up Ken's face. His eyes were closed and his face had a strange pallid look. He looked dead, and only John's first aid training enabled him to push away the sudden fear that he was.

'Perhaps I will sleep,' Ken spoke again, drowsily. 'Funny thing is, I can't feel my legs. Perhaps I wouldn't be too good walking . . .'

It seemed to take hours to get to the flat place at Shepherd's Ghyll . . . hours in the rain and wind, with the weight of a man and a stretcher bearing down on the straps.

John took his turn, taking the strap over his shoulder.

68

Once more, those hours of exercises in the rain proved their worth. It seemed almost impossible that they could get down this near-precipitous slope without someone losing their footing . . . but at last, miraculously, it was done.

Bill had contacted Geoff at the base of the fell and he in turn, through Jim Hodgkin in the relay Land-Rover, had contacted RAF Boulmer. The chopper was on its way. They staked out a flat central section on top of Shepherd's Ghyll and rigged up lights. Then they stood, buffeted by the rain, close to exhaustion themselves after the difficult descent, sipping hot tea, straining to hear the first sounds of the helicopter's approach.

Tony Moore was feeling Ken's pulse. He had lapsed once again into semi-consciousness. His skin was clammy and white. They were all anxious. It seemed ridiculous . . . in June, when people were probably walking about on Morecambe Sands under the lights, in their T-shirts. It seemed ridiculous that a man might die up here . . . from the wet, from the cold, from the sheer fear and trauma of injury.

'It would be a good thing to give him a blast with the Reviva,' Tony muttered. They fitted the mouthpiece of the warm air machine to the casualty's lips, letting him breathe the warm air which in turn would warm his inner core. 'His pulse is rather erratic, Skip.'

Bill nodded.

'Won't be long now.'

Bess came sidling up and lay down beside the stretcher. Bob looked at John and made a face. Bess had an instinct. It was one of the things that made her a good dog.

'I asked for a doctor to be on board,' Bill said, taking a sip of tea.

Tony nodded, busy trying to adjust the 'cas bag' to give the casualty maximum warmth without moving him at all.

'Good,' he answered eventually. 'I think it might be touch and go.'

'No bobble hats, I hope, lads.' Bill raised his voice against the wind. 'We don't want to get this far only to bring chopper down.' They nodded, only too aware of the dangers of stray objects being sucked into the powerful air intake. 'And tighten the straps on the safety helmets. It's all too easy to forget, tha knows.'

'I can hear it, Skip.' Tony Moore looked up. 'Chopper's coming.' Bess had heard it, too. She stood up, her nose into the rain, whining with excitement. The noise grew, clattering against the crag. Then the edge of the fell was lit up, blackly, by the powerful search-lights and the red flare, the rain scurrying away in the beam. The next moment it was hovering above them, tearing at their clothes, the sound seeming to shatter their eardrums after the quiet. The winchman was lowered, the stretcher secured, and they watched the two of them swinging up into the dark, turning and turning. Then the helicopter seemed to roll on its side like a great whale and to fall away between the crags, the noise fading. They were left in silence.

'Bloody good pilot,' Bill commented, his mouth full of Kendal Mint Cake.

'Hope he makes it,' said someone else in the dark. They all knew what he meant.

'Coom on, lads.' It was Bill's grainy voice again, stirring them into action. 'I could do with a gey good nosh-oop back at base. There's nowt but an empty space where me stomach used ter be.'

They set off down the fell, Bill reporting a successful lift-off to Control as they went.

Much later, when all the equipment had been stowed and they had been fortified back at base, during the de-briefing, with baked beans, bread and soup (courtesy of the WRVS), John drove home through the dark. The rain had stopped. The streets were deserted. He looked at his watch. It was three a.m. They would be operating on Ken Cullom in Newcastle General. It would be touch and go, just as Bill had said. Now, the adrenalin ebbing away, he felt terribly, terribly tired. It was work in the morning.

He opened the door with his own key. Sam came waggling down the hall, his face reproachful. He peeled off his wet socks at the door, and tiptoed in. There was a light in the kitchen. Tina was sitting asleep in front of the stove. She woke up with a start.

'Is he all right?' she said. 'Did you find him?'

'Yes, we found him.' He made his face smile. God, he was tired. 'But we don't know if he'll be OK. We'll wait and see.'

'I hope so,' she said. 'Let's go to bed.'

7

Whenever they worked close to each other, John had watched Bob Irons' dog, Bess, with fascination. The way she 'quartered' the ground, working backwards and forwards in complete co-ordination with Bob. They seemed to have an amazing accord—almost telepathic. It seemed a shame that they would soon be moving to Marrowdale.

'How's she done so far, Bob?' he asked one evening in the pub. 'When we've been training, she's often the one to "find" the body. Has she ever done it on a real rescue?'

Bob laughed and shook his head. 'No, it's not often that glamorous. And of course we are usually only involved in a search when we are accompanied by a Graded Search Dog, which narrows the field a bit. Until she's graded herself, that is. All you can do most of the time is declare that the area of fell you have been assigned to search is clear. That's terribly important, and it's one of the main reasons we don't mess about with half-trained dogs. Bess has never *missed* a body in the fells but she's never actually been the one to *find*. She will quarter a large area and, because she's so well trained, the rest of the team can be confident nothing has been missed. So she has an unfaulted record. The important thing is not who finds the "body", but that he or she *is* found as soon as possible. That's what being in a team is all about. It's just as much those members who eliminate their bit of fell from the search who "find" in the end, as the member of the team who actually pinpoints the "body".' Bob paused and looked at John. He grinned.

'Mind you, when it does happen to you, it's great. And it must be especially good when a dog handler "finds". After all, whatever the theory, we're still in the business of convincing a percentage of people that dogs actually *can* be an asset—and a tremendous one—during a search. So the more finds the better.' He drained his glass. 'Thanks, John. It's your shout.'

John took the glasses up to the bar for a refill, and fought his way back.

'It's time we started with Sam, then,' said Bob. 'You're keen, aren't you?'

John laughed. 'I'd like a bit of guidance, Bob. I don't want to jump in at the deep end. You lot are enough to put anyone off. I know he's OK with stock now, but how would he shape up to SARDA training? I know it's pretty tough. I don't want to make a fool of myself. Especially being an off-comer,' he added wryly.

Bob guffawed into his beer.

'You'll get plenty of stick, John, let's face it, when the boys hear about it. And more because you're a stranger. But I agree—let's have a look at Sam. How about Sunday afternoon at the base of the fell where the track runs up from the Horse and Groom? It's a good place to run a dog. Why don't you ask Derek to come over? I'm sure he'd be glad to give a hand.'

John readily agreed, but as he drove home, the thought of Sam's face as he lay at Arthur Crabtree's feet, dejected and bored, and the memory of the trainer's last words to him—'It may be that he will never be totally reliable'—made him wonder what kind of fools he and Sam were about to make of themselves.

Derek and Bob were waiting for him at the bottom of Langenhow Fell—Derek with his battered Land-Rover, his Border Collie hanging over the tailgate. They greeted him laconically. John had a nasty feeling of being talked about. He opened the back of the Estate. Sam jumped out, sniffed the rain-laden afternoon air and cocked his leg against the wheel of the Land-Rover.

'Thanks, Sam.' Derek grinned and winked at John. His Border Collie went into a frenzy of barking over the tailgate, eyes luminous with hate. John began to feel better.

'Shall we just walk on up—I brought Lad here—see how they get on? He's only like that when he's defending summat. Nearly got my ear bitten off coomin down here. Some woman with a Yorkshire Terrier asked the way to Kirkoswald. Lad screaming in my ear. I was that embarrassed . . .' He undid the tailgate and let it down with a crash. 'Out, Lad.' The Border Collie, a fine black and white dog, jumped out and he and Sam circled each other warily, tails stiff and only wagging at the tips.

'Hedging their bets,' Derek laughed. 'Come on, Lad. Leave! Let's walk on up the fell.' Bob opened the door of his car and let Bess out. The other two dogs forgot their mutual hostility and started fawning round her. Then they were off, leaping through the heather, Bess, a flirtatious glint in her eye, leading the way.

Suddenly, John saw a small bunch of sheep huddled on the fellside ahead of them.

'Heel, Sam.' Sam stopped, looked round and then left his companions to come loping to John's side. The other two followed suit. They passed the sheep, each dog cleaving to its master. Sam's expression in any other circumstances, John decided, would have been described as sheepish. He couldn't get away fast enough! And the sheep, reassured, only found it necessary to trot a few yards before they settled down to graze contentedly. They sensed the dogs were not a threat.

John spent the best part of the afternoon on the fell with Derek and Bob. There was no attempt to do any search training: they just walked and chatted. Occasionally they would stop, call the dogs to heel, and sit on a granite outcrop, looking out over the valley below, bathed in the afternoon light, the distant shadow of the Pennines stalking beyond. For the first time, John became aware of how much Sam had grown up. He was quieter, more mature, even though he still had his endearingly batty side. And the joy of having a dog up here in the wild, quiet spaces of the fell, a dog as a companion, a friend, reliable, not likely to let you down . . . it was what he had always dreamt of. That was the essence of it: not the obedience trials, but the quiet fellowship of a well-trained dog. Each man and each dog, separate from every other man and dog, relying on each other.

When they came down at the end of the afternoon Derek said:

'I think he'll be all right. Would you like to bring him to the next training weekend? We'll see how he gets on there. It's at Hammerbank in October. If he shapes up OK and you're still keen, we can register him as a trainee. But don't do anything until then. We like you to start from scratch with us. It's easier that way—rather than having to undo lots of mistakes, which is difficult and not very good for the dog.'

John was thrilled. He went home full of high spirits, bursting to tell Tina. She was cooking on the range when he came in, looking pink and harassed, the children playing round her feet.

'Hello, John. How did it go?'

'Derek's asked me if I'd like to come to a training weekend over at Hammerbank—October some time . . . He seems to think Sam's got potential. Anyway, I mustn't do any training on my own till then, just in case I start him off wrong.'

'Oh, that's marvellous, John. I'm so glad.' She stepped over the pile of toys on the kitchen floor, and kissed him on the cheek.

'And Derek says that if he shapes up well at this weekend, or the next one, he can be registered as a trainee. They have about seven of these training weekends during the year.' He bent down and lifted Anna off the floor.

'It's a lot for you to cope with, Tina. Do you mind?'

She smiled up at him.

'As long as you babysit for me on my Institute nights, I can cope with the rest. How about we come and help at weekends? Matthew can "body" for Sam when you start training . . . as long as it isn't more than four yards away! I'm so glad he didn't disgrace himself today. I always knew he was something special.'

Sam came waddling over to her, a woolly giraffe in his mouth.

'Just for that, Sam, I'll save you a sausage. You deserve it. Just keep it up!'

8

It was an hour's drive to Hammerbank. John got up very early while it was still dark. He packed his gear, and made a flask of coffee. Tina came into the kitchen, half-asleep.

'Your sandwiches are in the fridge,' she said, yawning. 'I meant to get up and make you some breakfast. You should have woken me.'

He grinned at her. 'I heard you get up to Matthew in the night. You deserved a lie-in. I had two great bowls of muesli; don't worry.'

'Have a good time. Hope Sam behaves himself.' She yawned again. 'I suppose the tribe will be waking up in a minute. I'd better get dressed.'

He kissed her goodbye and tiptoed to the foot of the stairs, letting Sam, who was squeaking with excitement, out of the front door as quietly as he could.

The sun was just rising over Garth Fell. There was a smell of night rain and chrysanthemums in the air.

'I think it's going to be a good day, Sam.' He walked down the shining, morning path with Sam prancing ahead, jumping up at the gate and running back. He unlocked the car and Sam jumped up onto his special rug in the front seat, which John had spread out the night before. He looked up at the windows of the house, blank with the rising light of the morning.

John had never seen the lakes look so beautiful, the October sun picking out greens and golds against the grey water, the becks swollen with rain, mist lying in the valley bottoms and across the lakes, a pink and gold sky fading to blue, then sudden clouds passing across the peaks which caught fire again in the sun as the shadow passed. They were both keyed up; Sam imprinting his nose in a wet squidge on the steamy window and singing to himself; John his stomach tight with apprehension.

Bob Irons would be there and Derek, but he would not know anyone else. Most of them who were not from local teams would have travelled up the night before and would be staying at the Residential Centre at Hammerbank. They would have had a get-together last night. Perhaps he should have gone. He would at least have met a few people. But he had had an evening clinic it would have been hard to put off.

The day was supposed to begin at 9.00 a.m. and it was quarter to nine when John parked his car in the grounds of the hostel. He had arranged to meet Derek inside, and, leaving Sam, he made his way to the main building. There was an assortment of Land-Rovers, estate cars, Range Rovers and a Mini, all parked along the approach road. As he passed each one a canine face looked out at him from the rear window or, in one case, the driver's seat. Mostly Border Collies, but one Alsation that barked at him ferociously. He saw Derek's Land-Rover with Lad peering out at him from under the canvas. Knowing Lad's reputation for guarding the car, he did not stop to say 'Hello'. Even so, he got an earful of abuse as he walked past.

There was quite a crowd milling about outside the side door of the hotel, amid a motley collection of rucksacks. As John approached across the grass, Derek broke away from the group and came towards him.

'Come and meet Mike,' he said. 'He's our training officer. He'll be starting the trainees off this morning—I've told him all about you.'

Half an hour later, John found himself at the base of the fell surrounded by half a dozen Collies and a couple of Alsatians. There were four or five who had obviously been on a weekend before and who, compared to the absolute beginners, were exuding confidence. Mike came round to each of them in turn.

'Would you do some bodying for us first time round, John? Just tie Sam up by the barn over there, and if you don't mind, get yourself well hidden in that section of the fell there. I want to do some work first with the dogs who are coming up to their Novice grade. They're pretty experienced, so we set them fairly hard tasks. They'll be hoping to be invited to the Annual Grading Course this year and we'll be assessing them this weekend. It'll be good experience for you, too—seeing how the dogs really work. If you can get yourself behind one of the crags, so that you can see what's going on but can't be seen, it would be a help for you.'

John nodded and, taking the leather lead out of his pocket, called Sam, who was nose to tail with a rather wary bitch Collie.

'Stay, Sam.' He tied him up to a fence post at the side of the barn, the last of a jumble of buildings, a natural extension of the farmhouse which abutted the track. Then he walked up through the muddy field which was stone-walled on three sides. Only where it swept up to and melted into the fell, did the stone walls peter out.

'I suppose they pen the sheep down here,' he thought, 'but other times they can just wander up on to the fell at will.' There was plenty of evidence of sheep here—what the locals called 'trunlins', but the sheep were scattered aloft around the crags. It was pretty boggy down at the base. Already John could feel that creeping iciness of water invading his boots. He was glad to climb higher away from the wet, as the sheep must have been.

It was sunny down in the intake fields, but here, suddenly, as the fell became steeper and the granite crags loomed closer, the 'clag' came down, covering everything in fine, grey rain. It was very cold. The dog and handler had been banished out of sight behind the barn, but now, in any case, they would not see him 'go in'. He clambered sideways on the rock, and banged his shin hard as the slippery crag bit back. He swore. Up there, where there was a sort of overhang: that would be a good place. He scrambled on, sweating a little, aware that despite the last few weeks with the Mountain Rescue Team and long daily walks with Sam to get him fit (on Derek's instructions), he was still pretty out of practice. The move, and the change of job, and the demands of the growing family, meant that serious climbing had gone out of the window.

He got his second wind. God, it was higher than he had thought. This would give the dog a run for his money. He would get higher into the tops. He heaved himself up onto the overhang, boots scrabbling. In the mist, how easy it would be to find oneself suddenly disorientated . . . up here, with the clag seeping coldly across his mind; and it was not as though he were exhausted or injured, or simply frightened. How would it be for an inexperienced walker . . . wrongly dressed . . . stumbling about in this? The sun was probably shining down below, but he was completely cut off . . . only the odd shout, the distant barking of a dog (it sounded like Sam) drifting up to him.

Climbing over the back of the overhang, he found a small hollow and squeezed into it, glad at last to be protected a little from the weather. He was out of sight of most observers, except someone coming along the tops. It had started to rain—horizontal needles slicing across the fell—and with it a sharp wind. It drove straight into John's hollow . . . but there was no time now to move.

Then, as suddenly as it had come down, the clag lifted. Brilliant sunshine flooded the fell. A soft zephyr of a wind blew into the hollow. John stretched himself out on the rock. It was unbelievable, like some kind of conjuring trick.

There was certainly a lot of noise coming from below. He looked over the edge into the valley. He could see Fred

Jepson's farm, and the pond at the corner of the barn. There was a group of people standing round. Dogs running everywhere. The water of the pond was agitated white by some sort of waterfowl . . . they were being chased. By a dog! An awful realisation began to grow on him. Far, far, away, the surface of the pond was broken by a smooth golden head . . . swimming about in the flurried waters; a lunge forward, frantic quacking sounds carried up the fell. It was Sam! He was sure of it. He looked over towards the corner of the barn where Sam was supposed to be tied up, straining his eyes. There was no dog there!

He started to get to his feet. Then he saw the Collie immediately below him. Fortunately, neither the dog nor his handler had seen him—but he would have to sit tight. Oh, Sam! John crouched down into the hollow once more, in a fury of impotence. What he would do to that misbegotten tripehound when he got hold of him!

When he looked again, the pond was quiet, and Sam nowhere to be seen. John hoped fervently that he was somewhere out of sight, getting his come-uppance. Why had he come? Sam was an idiot. They would both be a laughing-stock—at the very least. One thing was certain: they would never be asked again.

The next twenty minutes, until John was 'found' by the Collie, were very long indeed. John had unintentionally made things more difficult by lying under the rock, so that the wind was blowing straight into the rock wall behind him. It had made hard work for the Collie. But she was a bright dog. Using air scent she had eventually worked her way behind the rock and zig-zagged into the wind until she isolated the spot where John was lying. The handler's face, red with exertion, appeared a few minutes later.

'See your dog's playing oop a bit down below.' It was a handler from another Cumbrian team with whom Langenhow had a friendly, if edgy, rivalry. It did not help. 'You'd best go on down and sort it out.'

John grinned weakly and began to scramble down the fell. Mike was waiting for him by the barn, a very wet and rather subdued Sam sitting beside him. Sam was wearing his shifty-eyed look.

'Sam!' John said sternly. The tip of Sam's tail wagged feebly.

'I've given him a good shaking,' Mike said grimly. 'I don't think he'll play that particular trick again. And you'll need a new lead!' He held up the remains of the leather lead, chewed neatly in two.

'I'm sorry, Mike. I don't know what came over him.'

'Boredom, I would guess,' said Mike wryly. 'But I'm more worried about the implications as far as his stock training is concerned.' He paused and looked down at Sam, who looked away guiltily, his tail stirring damply. 'I want to go through a few tests with all the new dogs. First of all, we'll get Fred down at the farm to put them through a fairly stringent stock test. Let's hope Sam does OK on that one.' Sam's tail wagged feebly in response. 'I'll just get this other lot started. Then I'll be with you.'

John squatted down beside Sam, whose coat, spiked with water, was giving off a strong smell of pond weed.

'You are a damn fool, Sam. You've probably blown the whole thing. What did you have to go and do a thing like that for?'

Sam's otter-like tail banged the grass. He lifted his eyes to his master's. They were filled with deep penitence. John was unconvinced.

'It's no good trying to get round me. One more black mark and we'll be out! Never mind your fatal charm. I've got faith in you, Sam. Don't let me down again.'

Mike came up to them.

'If you're ready, John . . . ?' Sam, and the other two new dogs—a young Collie, and a German Shepherd bitch—were taken through a gate into one of the intake fields. Fred was waiting for them. He farmed at the bottom of the fell, and had vast experience of training dogs. Most important of all, Mike explained, he had a sure instinct for a 'bad 'un'. The dogs were always brought to him for initial vetting, before they were even allowed on the fell.

'If you can get past Fred, then all you have to worry about is Bob Morton on the Annual Course,' grinned Mike. 'And that's another story!'

They followed Fred, who set off without a word down the field. John and Sam were to go first. Fred stopped by one of the gates. In the field John could see a large flock of Herdwicks, pulling at the autumn grass.

'Now then, young man.' Fred's disconcerting blue eyes fetched up, with some reluctance, against John's face. 'Let's see if your dog is going ter play silly buggers with me sheep. I want you to let him go, first off. Then I'm goin' ter watch you work, like. See if you can concentrate. Go on, then. What are you waitin' fer?'

John was sweating, although it was a cold day. After the duck pond, if they put a foot wrong, they would not stand a chance. But he need not have worried. Sam behaved faultlessly. Even Fred unbent enough, at the end, to lay his hand on the golden head and pronounce:

'Tha's got thyself a gey good dog there, young man. I wouldn't mind him mesen!' It was praise indeed. John just hoped they had done well enough to wipe away the blot of the pond . . . and that Sam had learnt his lesson!

The Collie followed on after Sam with no problems at all. But for the German Shepherd and his handler the stock test was an abrupt end to all their hopes. As soon as the dog saw the sheep, he ignored his handler's commands and bounded excitedly towards them, scattering them in a panic. There was chaos in the field for a few seconds until Fred himself waded into the fray and dragged the reluctant dog out by the scruff.

'It's a good thing there's nowt here but young tups. Any other road tha dog would have yows abortin' all over place!'

The handler was white-faced with fury and disappointment. John was doubly grateful for the lesson that Sam had learnt at Greystones Farm.

Mike took the other two up the side of the fell, after getting a nod from Fred. John watched while the Collie was put through a basic obedience test. John was impressed by her performance. The dog was faultless.

'Right, John. I want you to put Sam through his paces, so I can see what sort of control you have . . . and how he responds to you. I want you to work up to a ten-minute stay with you out of sight. OK?'

Sam behaved perfectly. It was as though he, too, was trying to make up for his earlier disgrace. John experienced that rare, exhilarating experience of working in perfect accord with his dog—as though they were one. And at the end, when he was given the command 'stay'—a crucial test after the morning's problems—Sam could not be faulted. Eyes shining, body rigid with excitement, he lay totally unmoving until John released him with his final command.

Next, Mike took them higher up the fell, to test their fitness. Now Sam, freed from the more formal disciplines but still under control, negotiated scree and slope under the training officer's watchful eye. John had trouble keeping up, and he began to appreciate how fit he would have to be to cope with these sort of conditions, hour after hour, often in rain and snow.

'Well, he's fit enough,' Mike pronounced eventually. 'I don't think he'll have too much trouble working up on the fell—though don't forget, after five hours or so searching, in a white-out, it is only a supremely fit dog who can retain his mental faculties. He could do with losing a bit of weight . . . and you.' He looked across at John with a grin. 'You'll have to get some fitness training in as well!' John grinned back, somewhat ruefully.

'I'll work on it,' he said.

Both the dogs had come through the fitness test, and Mike took them back into the intake fields where Sam and Laura, the Collie bitch, were fitted with their fluorescent orange jackets.

'The dogs will wear these whenever they are working,' Mike explained. 'That way they get to know pretty quickly when they are on duty. You'll see a great difference in the dog's attitude after a while. And of course, it's a good idea for your dog to show up clearly. You'd be surprised how well the average German Shepherd, for instance, blends into the fellside!'

'What about night work?' asked John, trying to avoid being licked to death by Sam as he struggled to fit him into his jacket.

'Well, we have these coolights . . .' Mike pulled a couple of greenish tubes out of his pocket. 'They fit onto the top of the jacket like this . . .' He demonstrated with Sam. 'If your dog will stop wriggling about long enough for me to show you! They glow in the dark and you can watch how your dog is quartering the fell, even, say, half a mile away. And you'll find out how important it is to see what ground he has covered . . .'

They began to work. Mike showed them how to hold onto their dogs while the volunteer bodies lay down a few metres away. Gradually they established the vital sequence of 'seek', 'find' and 'return', with Mike watching them carefully. The dogs were encouraged to 'go in' and 'find', rewarded with a titbit, and then to return to the handler. Very gradually, the distance was increased. There were frequent breaks to stop the dogs losing interest, but Sam was as keen as mustard and showed no signs of flagging. Laura, too, flung herself into the 'game' with tremendous enthusiasm, showing the marvellous work instinct which had been bred in her by generations of Scottish farmers before her mother came over the border to England. This was Terry's second dog and was able to give John's some useful tips, although this did not stop the development of a friendly rivalry between the two handlers.

Mike advised them that they must never lose the element of pleasure, of achievement, which was the key to the dog's willingness to work. Once he stopped doing it for love, he said, you were in dead trouble. John wondered how he would ever be able to extend that simple little sequence to encompass hours and hours of searching for a body on the fell, a body which, in all probability, your dog would never be the one to find. How did you keep the sense of achievement going then? It seemed a very big step indeed . . .

It was a good weekend. Sam had performed well, John could feel it, and, apart from pinching John's sandwiches out of his rucksack and sharing them with a willing Laura, he had not put up any more 'blacks'. On the Sunday afternoon they all assembled back at the Centre—a mixture of muddy boots, rucksacks, wet dogs and soup, everyone talking and laughing. John felt a bit shy, but just then Mike called him over to the table, where a number of the officials were sitting, poring over maps and lists.

'OK, John. Forgetting—hopefully forever—the incident of the pond,' Mike began. There were smiles all round. 'We feel, after observing Sam for most of the training sessions, that he shows real promise. He is enthusiastic, highly intelligent, and has bags of personality. We were very impressed when we watched him working. Even at this early stage he shows his independence—sometimes a bit too much . . .' John swallowed hard. 'But though this may sometimes lead him into trouble, by and large his general standard of obedience is good and, almost more important, his own self-discipline.' Mike smiled down at Sam, who had crept up to Mike's knee under the table and was intent on ingratiating himself as far as was caninely possible. 'We shall have to work on this fine balance between the dog taking orders and using his own initiative . . . It wouldn't make any of them the stars of an obedience class, but when it really matters, that's when the dog's own character comes through. And, don't forget, in a white-out on the fell at night, it will be your dog who has the right equipment.'

'So,' Mike looked round at the others, 'We're all in agreement. We like Sam. We think you may make a good partnership. You'll need to work pretty hard on those exercises, gradually extending time and distance as we showed you. If you're agreeable, and we haven't put you off, we'd like to include you in the Register of Trainee Handlers. See you at the next training weekend.'

John drove home in a daze. After that terrible beginning . . . . to be told that they might have a chance, that Sam was the sort of dog they were looking for. A rare breed, someone had said, afterwards . . . we're looking for something special. It was a good feeling . . .

9

There followed more than a year of hard training. Daily
work-outs with Sam, early in the morning before John
went to work and again in the evening. And, spaced out
through the year, seven training sessions with SARDA,
weekends when trainee dogs, Novice and Graded Search
Dogs from each area in SARDA—the Peaks, the Dales, the
North-East and the Lake District, together with SARDA
Wales—could get together in a controlled environment for
guidance and advice on training. After each training
exercise there would be more to learn, more to practise. It
was a slow business and often, when Sam was being
difficult, John wondered if they would ever get to that
mythical marvellous moment of being invited on the
Annual Grading Course—a moment when he would know
their goal was in sight.

As Bob Irons had moved away and joined another team
on the far side of the fells, John had to train almost totally
on his own. He often found it hard to get a 'body' to
volunteer . . . it was no joke lying out on the fell in the
middle of winter, hoping to be found by Sam who was
quite likely to lick you to death. John wondered if he would
run out of friends before the training period was over. He
noticed that their faces took on a glazed look when he
approached them on a Friday . . . after all, what a way to
spend a weekend! Twice a month, Derek would make the
trek from Grasmere or John would drive over there, taking
the family to visit Ivy, and they would have a full day's
training on the fells.

It was an unrelenting routine, never to be missed whatever the weather. The 'find' sequence, the 'search' phase, introduced the idea of hunting for a body, and last of all the 'blind search', when John himself had no idea where the body was. It was a strange feeling, having to rely on Sam's ability to find, where before he had some pre-knowledge of what to expect; now, suddenly, he was in the dog's world—with fewer senses and without the dog's heightened instincts. It was a crucial moment. And when Sam found the body after quartering the fell for twenty minutes, things were never the same again. The close, sympathetic, almost telepathic, relationship which had been developing between them suddenly became only a part of something greater. All at once, it became a partnership of equals.

But from the beginning there had been one small problem: how to get Sam to bark on command. It was essential that the dog should be able to indicate a 'find' as clearly as possible, especially in bad conditions; and, although there were other ways of indicating, it was a useful asset, especially during training, to have a dog that would 'speak' on command.

Sam was heavily into body language. When he found, he would lick the 'casualty' enthusiastically (not always appreciated by the hapless volunteer), scamper round a few times chomping his jaws with excitement, and then gallop back to John to communicate in his own inimitable way how exceedingly clever he was. This would involve tail wagging, body wagging, wuffling and snuffling, rolling on his back and other useful body signals which John understood well enough—but in a life and death situation where it was of prime importance to be clear about every aspect of the search, it would certainly be preferable to have a 'speaking' dog!

John started with the body. He primed him with choc drops and instructions on what to do if Sam barked. Sam

wuffled and chomped but not so much as a single 'woof'
escaped all afternoon. In the event, Sam found the choc
drops on a ledge, ate the lot, cellophane and all, and that
was that.

The answer came to John one evening. Sam had an
unreasonable aversion to glass doors, especially the frosted
variety. They confused him and, if he saw the blurred
shape of a person on the other side, he would go mad with
excitement and rage. John had stopped taking him into the
bank in Penrith for precisely that reason—it was so
embarrassing. Unfortunately, there was a particularly
hideous glass door between the kitchen and the scullery at
the house. John had been meaning to replace it . . . but
there were always more urgent jobs to be done. Now it
suddenly came in useful.

John discovered that if he stood on the far side of the
door and called to Sam, Sam would bark madly at him. He
tried it several times, each time shouting out, 'Speak, Sam!
Speak!' as Sam began to bark. It worked. After several days
Sam had got the message. John had only to say the magic
words for Sam to explode. After a week, they had calmed it
down to a dignified 'woof', but then Sam began to develop
his own variations. Having discovered an extra means of
communication, why waste it? 'Woof', repeated several
times, with measured pauses, was the official language:
Sam used that when he was on duty. But a puzzled, rather
gruff 'woof' meant, 'You've forgotten a part of the
game/my choc drop/me . . .' and a whole chain of excited
'woofs' meant, 'I've found this body/why do I have to keep
coming down to you when you can hear me perfectly
well?/Oh, all right, I'm coming!' It was a sound which was to
gladden John's heart in the future when it floated down to
him in the deep night after long hours of searching on the
fell.

It was all part of the new relationship they were building.
And, not for the first time, John had that strange sensation

that it was Sam who was doing the teaching, not the other way around.

After a little more than a year, Sam and John were invited to the Annual Grading Course in February. Sam was to be assessed for grading as a Novice Dog. The course was to be based at a hotel in Keswick which had long been associated with climbers—the Crow Park Hotel—and the assessments would take place on the fells around.

John was close enough in Garth How not to have to stay during the four-day course, but as he skidded his way to Keswick on the Saturday morning along icy roads, he wondered whether he had made a wise decision. The roads in the town were snarled up, even out of season, with traffic and a light covering of snow. He left Sam in the car and ran up the steps, his stomach somewhere in his boots. The walls of the hotel lobby were lined with sepia photographs of epic climbs; grim-faced climbers in flat caps, walrus moustaches bristling, glared down at him from the distant, illustrious past.

On the wall there was a list of locations for trainee dogs coming up for grading as Novices, for Novice dogs coming up for Search Dog status, and for a small number of trained Search Dogs who were coming up for their three-yearly regrading.

John saw that they would be working on the north slopes of Helvellyn, pretty exposed in this weather; but first there would be the dreaded stock test. John was fairly sure of Sam by now; he had been working on the fellside amidst a liberal scattering of sheep for long enough ... but everyone he met in SARDA had their horror stories of dogs being failed because they had that mysterious thing, 'an eye for sheep'. John was not too sure what that was. He hoped that Sam did not have it—because that would be the end.

He went back to the car, pushed Sam out of his habitual place in the driving seat, and drove out of Keswick to the stock testing area, a mile of two on up the fell. Light snow had begun to fall again, the flakes buzzing like sleepy insects against the screen, seeming harmless, but already forming miniature drifts by the roadside. The fells were above him. The car slipped a little, climbing. How terrible it would be to fail.

By the time John got there, most people had arrived at the testing ground and were leaning over the stone wall, watching.

'He's down on Alsatians all right!' A grim-faced handler came out of the field. 'Says my dog was eyeing the sheep. Never heard such rubbish. I've come on a two-hundred mile round trip! Now I've got to turn right round and go back home. All that work wasted.' He stalked off. One of the graduate handlers, whom John knew slightly, caught John's eye.

'It's hard enough,' he said, 'but what can SARDA do? We rely on co-operation from farmers when we're working dogs. It only needs one bad 'un getting in among sheep and we'll lose all good will. Still, poor devil, you have to sympathise.' John nodded, wondering miserably whether Sam's fate would be the same. Another hapless victim was meanwhile going through his paces, trying not to show his anxiety.

'Old Bob Morton has the best eye for a troublesome dog I've ever seen,' John's companion continued. 'I work Collies on my land, so I know a bit about dogs meself.' The dog was snuffling at the ground on the far side of the field, having been 'sent away'. The handler was having a bit of trouble getting him back. John's sympathies went out to him.

'There's a bitch in season. I'm sure of that. Dogs have been reet scatty all morning.'

John groaned inwardly. The dog and his handler, and a huge Yorkshireman whom John remembered for his friendship and humour on the weekend courses, were standing in the snow talking to Bob Morton, a tiny wizened figure in a balaclava which almost totally obscured his face. Suddenly the handler turned and, with his dog at his heel, came out through the gate.

'OK, Dusty?' John's companion called over. Dusty's face lit up with a grin and he gave the thumbs-up sign.

'Ay, but I had to kiss him behind ear and whisper sweet nothings before he'd do what I wanted!'

'And that were only Old Bob!' Some wag in the crowd shouted out. Dusty went off laughing. John heard his name called.

'Good luck!'

John smiled his thanks. He went through the field gate . . .

It was an agonising ten minutes. There was no question of Sam taking any notice of the sheep—he barely glanced at them. But when he picked up those delicious scents at the far end of the field with his super-efficient nose, he was blowed if he was coming back. John sweated, and called.

Bob, his balaclava obscuring all of his face now but a pair of gimlet eyes, watched impassively as the snow swirled around him. Eventually Sam sauntered back. Bob sent him into the sheep with John. Sam took no notice of the sheep, and this time, with the wind blowing up the field, was untempted by anything else, much to John's relief. The sheep, bored and cold, trotted a few unwilling yards and then settled down to mouth at the frozen grass. It was hardly a case of mutual attraction. John came back to find one of the officials talking to Bob. He was explaining that they had found that one of the graduate dogs coming up for regrading had just come into season. She had been the first to run in the field. It was bad luck! Bob pulled his balaclava down and sucked at his teeth. He looked at Sam. Then at John. It had hardly been a sparkling performance . . .

'Tha'll have to attend to a few things, young man,' he said eventually. 'No matter if it were bloody poodle wi' bows in her hair sitting on mountain of beef bones . . . dog's got to come to tha! And it dizna matter ower mitch what excuse

93

tha have . . .' There was an awkward pause. He shook his head. 'Tha'll have to do more work on him, tha knows.'

John found his voice. 'Yes, I know that, Mr Morton. He's pretty good on the fell. Ignores the sheep.'

'We . . e . . ll. Jest keep him oop ter scratch. That's all.' He waved his hand. John, confused, looked across at the SARDA official, who was standing beside Bob Morton. He winked and smiled.

'You're OK. See you on Helvellyn!'

John walked out of the field gate. He could hardly believe he had got through. Now there were just the four days of assessments . . . could anything be as nerve-racking as this?

They were sent high up on the slopes of Helvellyn for the first day's training, taking advantage of the conditions for some snow searches. John was grateful that on the last training weekend, in the Peak District, they had had an opportunity to practise snow skills. Dogs working in England would rarely, if ever, have to deal with a real avalanche but, nevertheless, there would be plenty of searches in deep snow, with all the special abilities these would require. And, in white-out conditions, the dogs would really come into their own—their handlers deprived of sight and hearing, and only the dogs' superb scenting ability to act as a guide.

John had taken his turn to 'body' on that last weekend course, and his sympathies went out to the volunteer bodies buried on Helvellyn on a day when the wind-chill factor was around -30C. Ironically, they were probably sweltering—the temperature in the snow chambers would be amazingly high—but when Sam 'found' on that first morning and scrabbled away at the snow to expose the 'body', she certainly looked rather relieved! Sam had taken only fifteen minutes to locate Lucy, one of the most faithful of the volunteers.

The next two 'finds' took a bit longer, but Sam, aware

that he was on trial, plunged across the snowy slopes, quartering the ground with no sign of flagging. It was only when they had finished for the day, and John had towelled Sam down and given him some water, that he fell instantly and soundly asleep, snoring loudly all the way back to the hotel, his paws twitching as he pursued phantom scents across the fell.

It was a tough four days for them both. The first day had set a cracking pace which, if anything, accelerated as the days went on. But although John's leg muscles shrieked in protest after hours and hours quartering a stubborn hillside, Sam seemed exhilarated by it all, as though it were some gigantic game which he was determined to win.

John could tell, although the faces of the assessors gave nothing away, that Sam was doing well. He had only missed one find, because the time had run out, but he had not 'missed' the body in the area he had covered . . . and otherwise Sam had 'found' on every search. If John had been of a more poetic turn of mind he would have said that here, on these snowy slopes, Sam . . . flowered! A peculiar enough species, but there it was. A scatty, mad-cap creature who nevertheless had somehow developed into this—a dog of extraordinary powers, who might one day save a life. It was almost unbelievable.

The last afternoon came. There would be a special tea laid on in the hotel. Dogs unfortunately could not be invited, but it was the dogs who would be on trial with their handlers, and some, undoubtedly, would fail. It would be very hard after so much training, but there it was. The stakes were very high.

Despite John's conviction about Sam, he was suffering an agony of nerves. None of the assessors had said anything to him . . . Perhaps, after all, he had got it all wrong. Bess had been made up to full Search Dog with three others—she had not missed a 'body' at all. He had been a damn fool to come; of course Bess would make it—she was steady, careful, conscientious. But Sam! The assessors were reading out the list of dogs which were to be made up to Novice Search Dog. John did not hear Sam's name. It was not in the list. A sick disappointment filled his throat. After all that work! He could not bear it. All that work! Then, suddenly, Sam's name, and his own. He could not take it in . . .

'. . . Sam and his handler John Brown, winners of the Shield for the best performance of a Novice Dog during assessment.'

He got up, the cup he was clutching rattling over in the saucer.

'Sam! Oh, Sam! You've come up trumps!' He walked forward in a daze to collect the shield. The silver head of an Alsatian was embossed in the centre, and the names of past winners inscribed around. Images of Sam flashed into his mind—Sam, a carnation in his mouth, swanking round a show ring; Sam, bedraggled after his flirtation with the ducks; Sam licking Derek's face after a successful find; and, finally, an enduring image of Sam bounding down the crag, that very morning, tail wagging, body vibrating with excitement, a speckling of snow on his muzzle, planting himself foursquare in front of John, and SHOUTING (there was no other word for it!), shouting that he had found the body, and that he was the best dog in the world! It seemed a moment for which he had been made. There was a long way to go . . . the next target would be full Search Dog . . . but for now, it was enough.

Sam as a puppy.
Courtesy John Brown

Practising for his misdemeanour at his first Annual Course? *Courtesy John Brown*

John and Sam begin their training on the crags. *Courtesy John Brown*

Novice Search Dog Sam with the SARDA Shield he won in 1980. *Courtesy Cumbrian Newspapers Group Ltd*

Search Dog Sam, in his scarlet SARDA coat. *Courtesy John Brown*

Sam finds the 'casualty' and homes in.   *Photo: Paul Fearn*

In his enthusiasm, he attempts to lick the 'casualty's' face.
Photo: Paul Fearn

Sam 'speaks' to John, so bringing him to the right spot.
Photo: Paul Fearn

Keswick MRT ambulance, ready with stretchers and other rescue equipment. *Photo: Paul Fearn*

A Keswick team member checks the map references.
Photo: Angela Locke

John listens to the 'crack' (the briefing) at Keswick MRT. Behind him are radio sets that will be carried by the handlers.
Photo: Angela Locke

Sam in his element, surveying his home territory.
Courtesy John Brown

When landing conditions are impossible, the casualty has to be winched up. Sea King helicopters, used for longer-range operations and search work, from 202 Squadron, RAF Boulmer. *Courtesy John Brown*

Using smoke to indicate wind direction for the helicopter coming in to pick up the casualty – Langdale/Ambleside team.   *Courtesy John Brown*

The Sea King helicopter lands safely, John and Sam wait in the foreground.   *Courtesy John Brown*

Young Tyan picks up tips from Sam on a Search exercise.
Courtesy John Brown

Collies and handlers prepare for a Search exercise.
Courtesy Roly Grayson

Sam with Tyan – full Search Dog with a hopeful trainee.
Courtesy John Brown

Still very much a family dog: Sam with Tina, the children
Matthew, Anna and Pippa, and Tyan.   *Courtesy John Brown*

10

The phone seemed to have been ringing and ringing. He crawled out of bed. There was an extension on the upstairs landing.

'Hello.' He stood on one foot on the bare boards.

'A SARDA call-out, John. They want all the Cumbrian dogs if possible over beyond Pike Fell. A couple lost up above Garkle Crags. Rendezvous at the base of Pike, down by the main road. Soon as you can.'

He was suddenly alert. Downstairs he could hear Sam, woken by the sound of the telephone, banging his tail against the wall as he waited to be let out. He tiptoed over to the window in the darkness of the bedroom, falling over something sharp on the floor. He parted the curtains. There had been snow for a solid week now and he could faintly see the gleam of falling flakes.

'God, what a night,' he thought. 'Why is it always foul when we get called out?' It was a rhetorical question. He switched on the bedside light. It was ten to two. He put on all his layers one by one, hurrying, fingers fumbling a little. It was a long way to Garkle Crags, but mostly main roads, cleared daily now by snow-plough.

'Was that a call-out?' Tina asked sleepily. 'What a God-awful time.'

He leaned over and kissed her.

'They want the SARDA dogs. I don't know any details. See you soon, I hope.'

'Your equipment's in the cupboard downstairs. Everything's packed. I'll get up and make you some sandwiches, and do you a flask.'

'You're smashing. Thanks, love.'

It was the worst time of night. The snow-ploughs would not be out again till dawn and the roads were covered with a layer of icy snow. He needed all his concentration. Sam, wildly excited, leaned forward in the passenger seat, his eyes on the road.

He swung across onto the 'main' road up to Garkle Crags. It was hardly more than a country lane at the best of times. Now, with high drifts of snow on either side, it was a single track, and dangerous. John drove on carefully, shifting the gears to get maximum traction as the road began to rise. He leaned backwards and unclipped his radio from the side of his rucksack. With the snow drifts it was unlikely he would pick anything up, but still . . .

'Control from Search Dog John. Are you reading me? Over.'

The radio 'squelched' quietly to itself, emphasising his sense of aloneness. He left it on. The car moaned to the top of the pass and suddenly there was a burst of cross-talk crackling over the radio, and simultaneously he saw the Land-Rover lights below. The car slithered downwards towards Garkle Crags . . .

The search was being co-ordinated by the Peldale/Marrowdale Mountain Rescue Team who had been called in initially by the local police. They had two Search Dogs of their own attached to the team. Since the alert, at 19.23 hours, Bob Irons and Bess had been searching a large area of the south side of Garkle Crags while the rest of the team conducted a line search on the north side. They had drawn a blank. At that point the other dogs had been called in.

The Co-ordinator explained that they had minimal information to go on. A car had been found abandoned on a minor road between Garkle Crags and Peldale Pass, which was partially blocked. It had been identified as belonging to a couple who lived in Peldale—the Morrisons. No one could think of any reason why they should be travelling except that they were thought to have relatives in Danebeck, a village further up the valley.

The police had found faint tracks, now almost obscured by snow, leading away from the car towards a short cut which was often used in summer, over the lower end of Garkle Crags. After that the tracks had become confused. Bob Irons, whose dog Bess was a fairly new member of the Marrowdale team since his move last year, had had difficulty convincing the new police sergeant on duty that 'search' dogs were not 'tracker' dogs and would not be able to follow a trail across the fell. It was the difference between ground scent, and the air scent which Search Dogs used.

The M.R. Team had spent almost seven hours covering the vast snow-covered fell around the car, with no success. The Peldale Co-ordinator had nevertheless resisted calling in extra SARDA members until the whole of the north side had been covered. Snow had fallen throughout and it was dark, and any additional dog searches would now be hampered by the number of people who had been in the area already. Bob Irons and Bess were sitting in the Land-Rover, exhausted and wet.

By now four other Search Dogs from the Lake District area had arrived at the scene. The Peldale Co-ordinator was answering questions while plans were being laid for a major sweep search. This would involve men and women from the MRTs walking up all the fells in line formation, with possibly helicopters overhead.

No one had seen the Morrisons leave the village or knew what they were wearing, so they could have been wandering about in the snow since morning in inadequate clothing. They were known to be in their fifties. The police report had said that the car engine was quite cold. The outlook began to look very bleak indeed. The police had been unable to find anyone in Danebeck who knew the Morrisons, but a phone call had been logged at 17.15 hours, which eventually led them to a farm a mile or two up the far fell, where Mrs Morrison's cousin had been expecting her guests since lunchtime, growing increasingly worried. Had the Morrisons, panicking in the snow, left the car to look for help and, as they snow begun to fall in earnest, lost their way among the fells?

The second phase of the search began. John and Sam, as the Novice Dog and handler in the SARDA team, were given an area already covered on the north face of Garkle Crags, adjacent to Bob so that they could keep in radio contact. The conditions were very bad. The snow roared across them in waves, the wind threatening to pull them off the fell. The dogs, equipped with their orange jackets with the coolights glowing in the dark, moved away ahead of them as they climbed the fell path. Bob, after nine hours, looked exhausted, but he reckoned that the change of clothes, some brandy and hot soup had set him up at least until dawn.

They parted at the head of the track, sending their dogs away into the wall of snow, working obliquely into the wind. The dogs clawed their way through the gale-driven snow, as though it were solid. Sam and John took the upper part of the fell. John tried to visualise the area in his mind from the map—the crags and crevices. He would have to work each area minutely, trying to allow for changes in wind direction. It seemed an almost impossible task.

Sam was suddenly there beside him in the screaming wind, the feel of his coat under John's hand infinitely comforting. John screwed up his eyes against the blinding snow, trying to think.

'Away, Sam, away. Find.'

He threw up his arm, but Sam, following instincts of his own, bounded off slightly to the left.

John, much occupied with scrambling through the snow-caked scree, was grateful for the crampons he had fitted before they left. He hoped Sam would come back, blown back to his master; a small frail creature battling alone across the fell. All that training—thank God for it now.

'Search Dog John from Search Dog Bob. Are you OK?'

'. . . Fine . . . I'll keep to the area above you, Bob, and work up towards the Pike and over it.'

'. . . Roger . . . Over and out.'

The radio returned to its 'squelch'—a strange sound like phantom footsteps in the snow. Once more, in the dark, he felt its comfort. There were sudden moments of white-out, the snow hurling itself at him, knocking him sideways. Then the wind would drop, the snow dwindle to a few flakes, and he would be able to see his feet, and the fellside lit up in front of him by his torch. Then, just as he had got his bearings, the snow would pounce again, leaving him breathless. Where was Sam? He plodded on.

The snow cleared, and suddenly the fell was exposed in its white silence by a drifting moon, which peered briefly out of the clouds. John looked upwards to where the bulk of the Pike, hanging over the north side of Garkle Crags, reared up into the black sky. On either side, the white tresses of snowfields cascaded down, smooth, untrodden.

'God, I've got to get up there,' he thought. 'They won't have come this far but I'll have to get Sam to check it.' He saw Sam's small green light, like a distant glow-worm, quartering backwards and forwards under the crag. Sam finished on the far, eastern boundary, by the wall and John whistled him back. Sam came carefully through the scree, his tail wagging. John pulled his ears and spoke to him.

'Away, Sam, away. Find.' He sent him off at an angle towards the most distant snowfield. They would never have come up this far . . . but still . . . Sam's small brave light zig-zagged away. With a roar the snow leaped on him again and he was knocked blind. It was like a wild creature, plucking at him. He pressed the switch on his radio.

'Search Dog Bob from Dog John. Do you read? Over.'

There was the usual pause. Then Bob came through.

'. . . I'm covering the far flank up to the wall, as agreed. It's bloody awful up here. Drifting. Otherwise OK. Nothing yet . . . over and out.'

He struggled on through the white blanket of the wind. Sam came back and he sent him higher still, working upwards obliquely towards the blackness of the Pike which sometimes appeared during lulls in the storm. As they got closer to the arête on the eastern side of the fell, which was crowned by a crumbling wall, the going got much tougher. The wind was causing serious drifting in the deep gullies, and they had to find a way round each one, working their way back down again on the far side.

The Pike grew closer. Suddenly, with the capricious nature of weather in the fells, the snow clouds were whipped away, the wind died completely and the moon appeared once more in the torn curtain of the sky, riding silently over the fell. John could see Sam quite clearly quartering the rocks at the edge of the snowfield. He called him back, anxious that the dog might plunge into the new-laid whiteness and sink. It would be treacherous to cross until there had been a frost or two.

Sam came to him, a little reluctantly. John made a fuss of him, feeling for injuries. The dog was excited—John could feel it through his coat—and when he was sent away again he bounced off, looking back after a few yards and then bounding on. John had sent him into the area of igneous rock between the two snowfields which, despite the blizzard, still lay black and bare, stripped by the wind, wicked granite stumps rearing up like a dead forest. But Sam veered once more to the right, towards the snowfield. John began to run on the slippery scree, cursing, using his whistle to recall the dog. He could see the light moving. Then, clear as a bell in the cold air, a distinct bark. Once.

Twice. The back of his neck tingled. Not there, surely not there!

He could not see Sam anywhere. Had he gone out on the snowfield? John was sweating with exertion. He was still five minutes away from the base of the Pike.

God, don't say he's gone onto the snowfield. He got a scent of something. He'll be mad with excitement. John blew his whistle, hardly having enough breath left to call.

'Sam! Here, Sam!' A blast of snow hit him out of a clear sky and everything was blotted out. He lost all sense of direction, seemed to be whirling round in a swarm of wild black flakes. There was barking close by. The wind died as suddenly as it had come. But this time, the snow went on falling out of a clear, empty sky lit by the moon. It was uncanny. Sam was in front of him, the snow on his shoulders lit ghastly green. Sam was trying to tell him something.

'Woof!'

It was the clear language again. There was no mistaking it. And that familiar chomping of the jaws which he had never been able to give up.

'Show me, Sam! Show me!' Sam wheeled away and, looking back every few yards, he bounded off towards the base of the Pike. John lost him again. Then that faint blur of green against the shadow of the rock, close to the snowfield.

'Woof! Woof!'

'All right, Sam. I'm coming as fast as I can.' He gasped. Damn the scree. Easy to break your ankle. Slow down. Another few seconds won't make that much difference—or will it?

He reached the base of the Pike, the moon suddenly blotted out in the black shadow by the rearing mass of rock. Sam came back to him a third time, indicating clearly. The snow had stopped. They stumbled through the moonscape at the crag's base towards the eastern snowfield—what

103

should have been a gully, before the snows. Very deep, if he remembered the map correctly. Sam, rigid with excitement, was trying once again to get out on the snow. John tested it with his foot. Without a frost to harden it you could go down and down. It stretched away from them to the far rocks, under the moon, untouched, unmarked. They could not be under there! It was not logical. An elderly couple, this high up!

'Search Dog Bob from Search Dog John. How do you read? Over.'

'. . . Reading you strength five . . .'

'I think Sam's found something, Bob. Can you climb up to us? I'm coming down to the bottom of the gully on the east side of the Pike. Then we're going to work our way up the other side.'

'Dog John from Dog Bob. Roger. Will rendezvous on the south side of the snowfield. Listening out!'

John and Sam worked their way carefully down the side of the gully, John testing with the shaft of his ice axe to see where the snow would bear his weight. As they started up the other side, Bess came running over the crag, the green coolight glowing on her back. She and Sam bounded off up the right-hand side. John waited while Bob caught him up. The snow had stopped again and it was once more very clear.

'What's up, John?' Bob was very out of breath.

'Sam's either found something in the gully or it's on this side. He indicated on the left-hand side of the snowfield. I thought we'd both better look here. If it's in the middle, it'll be tricky.'

On the clear air came the distinct sound of two dogs barking.

'That's it,' said Bob. 'Let's go on up. There's something there.'

As they began to climb carefully, testing the snow on the edge of the gully, the dogs bounded down to them in a flurry of snow, both desperately excited. Bess jumped up at her master, putting her paws on his stomach and woofing in his face. It was her way of indicating. Sam chomped his jaws.

'Show me, Sam!'

'Come on Bess, then. Show me!' The dogs began to run back . . .

*　　　*　　　*

Meanwhile others were searching, too . . .

Joe Broughton and his dog Ruff lived on an isolated farm half-way along Hadrian's Wall. It had taken them almost two hours to get to Garkle Crags in the Land-Rover and by then most of the SARDA dogs were already on the hill. Joe was given an area in the farthest corner of the search area, away to the south-west of the road to Danebeck. He would have to drive up the Pass as far as he could and then leg it up the fellside from there. Joe had been up most of the previous night with a sick ewe. He was not in the best of tempers.

He opened the cab of the Land-Rover, fighting a vicious gust of wind, and swore. Ruff jumped down from the passenger seat, and together they started through the snow-laden bracken. Ruff, a black and white Collie, running round and round him in circles, barking with excitement.

He strode on up the fell, shining his torch ahead of him to get his bearings. It was an area not familiar to him. He was more used to his own patch, the hills of Northumberland. Still, years of farming his own sheep across the bleak moorland around the Wall had given him a feel for land—he seemed to have a sixth sense for where there might be hidden gullies, traps for the unwary in the snow, and he had memorised the area pretty well back at Control. He would keep to the south side of the fell, close to the road and following the small stream—the Danebeck—which had carved a gully for itself in the fold of the hill. From there, he would turn up the track towards Bartle Tarn.

He tramped on, sending Ruff ahead of him into the wind. Of course, they would not be this far up—or this high—but he would have to check. It was a bitter, black night, shuddering occasionally with snow. Here, in the curve of the fell, they were protected a little from the main blasts of the blizzard, but it was bad enough.

They climbed higher. Joe pressed the switch on his radio and gave his position. The snow was falling harder now, half-blinding him. He hoped it would not get too deep. It was always a worry with Ruff's long hair—that it would ball up in the snow. The only problem with Collies, which were otherwise so ideally suited for work on the hill.

The dog came back to him, asking for instructions. He sent him higher still, up towards Bartle Tarn. There was a track that way leading over to Danebeck village. It was always worth a try.

He pressed the switch on the radio.

'. . . Control from Dog Joe. Any luck down your way? Over.'

The radio 'squelched' to itself for a few seconds. Then . . .

'Search Dog Joe from Control. We've nothing yet. The weather seems to be worsening. We may have to call a halt if it gets too bad . . . How is it with you?'

'. . . Fair to middlin', I'd say. We can go on for a bit.' A gust of wind, weevilling its way through a crack in the fell, nearly knocked him off his feet.

'. . . OK, Joe. Keep us posted. Listening out.'

The radio went back to its 'squelch'. They were high up now, close to the ridge which hid Bartle Tarn—a small, dark pool which the locals had surrounded over the centuries with a host of ghostly stories. Joe remembered having walked the fell path as a boy, with his father, Ruff's grandmother at their heels, on a rare day's holiday away from the farm. Then he had been frightened by the strange, still tarn with its steep, rocky sides. He shivered involuntarily. It was the deepest part of the night, just before dawn. A ragged moon riding in the sky. A good thing he had grown out of that nonsense years ago.

He stiffened. There was a sudden barking over to his left. He listened in disbelief. The snow was falling thickly, deadening sound. He shone his torch. Nothing but a cone of light, snow slipping into it silently, then out into the dark. He couldn't see a bloody thing.

The barking came again. Definitely over to his left. He would swear that was an indication. He stumbled off the thin sheep track he was following, towards the sound. Then Ruff burst into the cone of torchlight, moving in, eyes luminous in the light. He was barking fit to bust, his tail rigid in that way he had when he was 'speaking'.

'Show me, Ruff. Show me!'

The dog ran off obliquely up the fell. Joe, floundering through the snow, did his best to follow. The bracken had caught the snow in waves here, like a frozen sea. On every other footstep the whiteness came up to his waist. It was hard going.

They came to the little plateau above the ridge, but they could have been in the middle of Morecambe Bay for all he could see—except it would have been a darned sight easier on the feet. The snow here, without protection from the fell, streaked horizontally. It smacked him straight in the face, making him gasp. Ruff was in a frenzy—running backwards and forwards, the coolight on his back weaving lines of light like a child's sparkler on bonfire night. Joe was bemused by the cold, struggled to clear his wits. There was something. His torch picked it up—two bundles huddled by the path which led away from the tarn towards Daneback. He fought his way over to them, spitting out snow, constantly wiping his goggles.

They were dead. Just, he thought. Lying in the lee of the tarn rocks, the woman in a tweed coat, her legs in stout shoes and stockings twisted oddly as though she had fallen—and he, lying down beside her, had placed his own coat over them both. He swallowed down sadness and sickness. They were not the first dead bodies he had found on the fell. Nor Ruff, who now, his job done, lay nose down in the snow, whining to himself in quiet grief. It never got any easier. And that sensible coat now half-covered in snow and the rigid face lit by the torch. He swore viciously to himself. What a sad, stupid, bloody waste. It always was. And how did they get this high up? There was never any telling what folk would do, mazed by cold and weather and panic, blindly following a path.

He got on the radio.

'Control from Search Dog Joe. Have located two foxtrots.' That was their tactful code for a dead body. It did not take away the image, though. You never forgot . . . He gave his position and then, squatting down beside the bodies, Ruff whining deep in his throat as he watched him, he shone his torch on the still faces for the last time. Very gently, taking Mr Morrison's overcoat and brushing the snow from its folds, he covered them both . . .

*   *   *

. . . The radio crackled.

Dog John from Peldale Control . . . We have located two foxtrots on the eastern approach to Bartle Tarn. Casualties located by Search Dog Joe. Team are organising stretcher party. Many thanks for your efforts. Rendezvous at Garkle Crags. Over.'

'That's miles away!' Bob exclaimed. 'Quick, get on the radio. I'm damned sure there's something else here. We'd better tell them.'

John pressed the switch.

'Control from Dog John . . . we have two clear indications from Search Dog John and Search Dog Bob. We think there's something here. Will investigate and report back.'

'. . . Roger. Will send up other SARDA team members to give you a hand. Over.'

John described where they were.

'. . . We think there may be something on the south side of the crag. Going on up now. Listening out!'

The dogs had disappeared into the shadow of the hill above them, but almost immediately the frenzied barking began again, each dog's voice quite distinct.

'They've found, all right,' said Bob. 'Let's get a move on.'

The dogs were leading them up towards a small crag which hung over the litter of rocks below. First Bess and then Sam came back and spoke again, impatiently. They clambered over the rocks which were piled high with snow on their north faces, making them treacherous in the dark. Once John slipped, his boot jamming into a vice-like slit in the rock, and he swore with pain.

The dogs had disappeared into the darkness at the base of the crag. John and Bob climbed in after them. Deeper in. It was almost a cave, enclosed from above and behind by the overhanging crag, and with a small gully running at right angles to the hill. There was very little snow here, and they were out of the wind. The dogs were going mad, the sound of their barking echoing sharply round the frosty fell. They were half-deafened by it in the narrow confines of the gully.

'Show me, Sam! Show me!' John slipped again on the crag, sensing the urgency in the dog's voice. It was dark in there, cut off from the moon, despite their torches. Then they saw the dogs clearly. They were half-way down the gully where their coolights glowed dimly on distant ice—a frozen waterfall caught for a second: green daggers of light. Then the torches caught the other thing in the gully.

109

A bundle. The two dogs standing close over it for a second, then once more racing down the gully, then back, excited, concerned, urgent. The men climbed down. It was a haversack.

A few yards beyond there was a body. It was wrapped in a polybag and it was very still. Sam was licking its face. Did that mean it was alive? John got on the radio. The rest of the SARDA members were making their way up the crag. He gave their position and reported to Control that they had found—a casualty—condition as yet unknown. Bob was examining the prone little shape, unrolling the bag, looking for a pulse. The dogs had quietened down; were lying close, protectively.

'He obviously knew what he was doing,' Bob said. He crouched over the body, feeling gently—for injuries, for small signs of life. 'Had all the right equipment.'

'I suppose he was benighted on the fell, and tried to get out of the wind. No wonder the sweep search didn't spot him . . .' There was a pause.

'Is he alive?' John asked.

Bob nodded. 'Just. We'd better get him into a bivvy sack with one of the dogs as soon as we can.'

Between them they lifted the inert body, wrapping it in John's duvet jacket for the time being, the feet in his rucksack which he emptied onto the rock.

The lights of the other SARDA members could be seen weaving up the fell, and one by one the Search Dogs, picking up the scent of three people, arrived in the gully, 'woofed' a couple of times and then ran off to indicate the whereabouts of the party.

More information was coming in over the radio about the other casualties. Then Bob Irons gave Control a full report on the condition of their own casualty. Control dispatched the doctor from where he was attending to the bodies of the Morrisons. Bob and John had been unable to find any injuries in their brief examination. It rather seemed that

110

the young man, benighted and caught in a white-out on the fell, had taken the very sensible step of getting somewhere out of the wind and making himself as warm as possible. But the fact that he was now unconscious with hypothermia not only reflected the extreme cold, but also might indicate that he had been there a long time. Either way, it would be touch and go.

By now they had been able to wrap the casualty in a bivvy sack. John eased himself in, and Sam, after floundering about a bit, not really understanding what was required of him, eventually ended up on John's chest, though not before John had experienced several mouthfuls of wet paw. Immediately the heat generated by the bodies—Sam's higher body heat was particularly useful—began to radiate around the bag. Hopefully it was providing the right sort of gentle warmth that might save a person in the last critical stages of hypothermia. Any direct heat, by making the blood rush to the extremities, could precipitate a fatal heart attack.

Someone had produced a stove, and tea was on the way. The wind had returned, prowling round the fell above them but unable to reach the hollow. But it was now close to dawn and the cold was intense. Men who had been walking hard all night now themselves began to feel chilled. John lay in the sack with Sam, feeling the cold slab of the body next to him—was there any warmth there, any life? Every few minutes they would check the young man's pulse. It still fluttered with distant life, but grew no stronger.

The doctor arrived with more specialist equipment, as the first faint grey light began to seep into the sky. It included the Reviva, which they played over the casualty's mouth, warming the air. John, close to him, could feel the slight rise and fall of his ribs, like a faint, last sigh, but there was nothing else. He was terribly, painfully young. They could see that now by the light of the torches and in the glow of the primus which gave colour where there was none. No more than fifteen or sixteen—no beard— but nonetheless sensible, well-equipped . . . if it was ever sensible to walk alone in winter on a fell . . .

'We're taking the Morrisons down by stretcher,' the doctor said. 'The helicopter will do for this one.'

John was terribly tired. In the warmth of the bivvy sack he could happily have gone to sleep. Sam, having done his bit, was snoring loudly on his chest. Bob Irons and Bess, themselves in a state of exhaustion, had gone down the fell, but the other SARDA members, who would be needed to help with the stretcher, were left behind with the doctor and a small 'crash' party which had arrived with the extra equipment. There was much discussion on the radio. The gully was enclosed and there was nowhere nearby suitable for a helicopter to come in, even if it did not have to land. They would have to move the young man, but in his present critical state the trauma might be fatal. They must wait. The rest of the team had set up a tent over John and the casualty. There was a comfortable smell of paraffin and tea; the occasional joke. It might well be they would have to stay all day—while there was life in the quiet, senseless body. Afterwards, if they were unlucky, it would not matter so much.

After a couple of hours the doctor, who was drinking tea with his knees drawn up, suggested they had better dig themselves in for a while.

'Is there any improvement?' John asked.

The doctor shook his head. 'He's warmer. A bit warmer. But that's about all I can say. It's about time you came out now. Change with someone else.'

John nodded, but as he made ready to ease himself out of the sack without disturbing the body, he felt a change.

He stopped moving.

'Hang on a second, Doc. I think something's happening. Come and have a look.'

The doctor crawled over the prone bodies and dogs in the

112

tent, the orange light of full morning taking colour from his face.

'What is it?'

'It's his breathing. It's different.'

Doc got out his stethoscope.

'You're right. He's breathing very deeply now. I think he's coming round.' He leaned over and looked at the eyelids; the mouth. 'Lost that bloody awful blue look, though it's hard to tell in here . . . He may make it. Come out now, John. We've still got to be very gentle, but we can get him warmed up now a bit more conventionally. John, will you get on to Control? Tell them he can be lifted off in . . . say . . . three hours. I'd better have a word, too. Tell them to get a bed ready in Intensive Care at Newcastle. He'll be bloody lucky if he makes it. Just keep our fingers crossed . . .'

Three hours later the still unconscious body, now wrapped in a fleece-lined casualty bag and strapped to a stretcher, was lowered down the snowslope with great care, to where the fell opened up, away from the base of the Pike. The helicopter was unable to land anywhere in the snow. They stood in the wind flurries from the rotors, watching the winchman swing down, and then the body turning and turning as it was hoisted into the chopper. It was after midday. There was a bright sun and a bitter, frosty wind.

John's eyes were gritty from lack of sleep. Although he had been comfortable enough—cocooned in a bivvy sack while the others had hunched uncomfortably on the rock, he felt as though he had given everything, as though, while he was lying there with Sam snoring obliviously on his chest, he had been willing that cold, slab of body to live. As though the life, ebbing away in a whisper, had somehow been called back by his own will.

Doc said it would be touch and go even now. That another few minutes and it would have been far too late.

That only Sam and Bess, scenting him out, had saved him. That the young man who, even now, might not make it, but who had at least a chance, would have lain in his unintended grave, eventually covered by wind-blown snow . . . perhaps never to be found. It was a sobering thought. And if it had not been for the Morrisons . . .

*   *   *

There wasn't always a feedback. People whose lives were saved on the fell were not always keen to contact their rescuers. Often, if they had got into trouble through their own foolishness, plain embarrassment made them want to forget the whole thing as quickly as possible. For some, coming so close to death was something they simply blocked out of their minds. But just sometimes . . .

Four months later, when the Peldale team had collected together at HQ for an exercise, a young man knocked on the door. He came in; self-consciously. In the envelope which he handed over was a cheque. He had been back on the fell. This time the snow had gone and he had walked with a friend. They had taken a week to walk the major peaks in the Lake District, sponsored by their village, their church, their school. Between them they had raised almost £300. It was Simon Parker's way of saying thank you.

11

That first year was amazing. Apart from the thirty call-outs in which the Langenhow team were involved, there were six separate SARDA call-outs. By the time Christmas approached, John began to feel that he had seen all human life out there on the fellside and that nothing could surprise him any more. There had been fear and courage, rank terror, hysteria, death and stomach-turning injuries, rescues which turned the heart over and made one feel good, rescues where such blind stupidity was involved that John had driven home in a rage, crashing the gears on the hills, infuriated by the potential waste of life.

But it was often Sam who, when things looked grim, brought humour and warmth as a gift to them all. Life was never dull when he was around: whether it was the sudden sloppy tongue on the side of John's face when he was snoozing on the sofa; or visitors' shoes watered with Sam's watering can which he carried lopsided in his mouth; or all the irritating, lovable, crazy little quirks of his personality, which made Sam so uniquely just . . . Sam.

And when John came down from the fell in bad weather, depressed because the news was bad, and yet another casualty had had to be winched off the crag, it was the warm, furry feeling of Sam close beside him which sometimes made it all bearable, and pushed the dreadful things away, out into the vast loneliness, to the farthest limits of the night.

But owning Sam was not always good for the nerves. Becoming a Search Dog had not changed his other, crackerdog personality. John sometimes thought he was worse . . . just to make up for being good and responsible for hours on end. And occasionally John wished that Sam was not quite so prone to causing maximum embarrassment with the minimum of effort. There was the awful experience of the car, for example. It took John a long time to get over that. Sam, however, seemed quite unmoved by the whole experience.

John often took Sam with him when he knew he would be going out on his 'rounds', visiting the elderly people who might be housebound and dependent on Social Services to keep an eye on them. Often they lived in small communities high up in the fells, where the husband might have been a shepherd all his life, before arthritis crippled him, or farmed a tenant farm before old age and ill-health forced the couple to leave the farmhouse and buy a bungalow, perhaps in a village where they were hardly known. They were proud people, independent and used to working hard, often until they were well into their seventies. It came hard for them to accept help at all.

On this particular day, unseasonably warm and dry for the end of March, John knew he had to drive right up to the top of the Pennines to a little village. As it was such a fine day they might get a chance of a walk. He would take Sam for company. And there was always the possibility of a call-out: John would leave his whereabouts with Tina and the office, in case Langenhow or SARDA needed him.

He had arranged his day so that the last call took him to the felltop village of Anston. Before they got there, John stopped the car and ate his sandwiches. He had not bothered to stop for lunch. As usual Sam had the last one. It was difficult to resist that special 'thin' face which Sam put on when he was on the scrounge. Anyone would think he had not been fed for weeks.

Then he let Sam out for a run around. It was wonderfully clear, with that bright metallic clearness you get in the fells

116

in early spring. Even the bracken fronds had tentatively put out the first shepherds' crooks of new shoots over the brown. It was very early for such signs of spring. The snows might come back. There was still snow on the tops, scenting coldness down into the valley. But today it was good to stretch and breathe in the sun.

Sam was having a marvellous time, and pretended deafness once or twice when John called him, eventually coming back with a muddy chest and legs after lying in a puddle to cool off. John dried him with an old towel he kept in the boot for just such a purpose, while Sam panted up at him, looking happy. He loved having his chest rubbed, even if he was being simultaneously scolded for getting dirty. The two usually went together.

John spread the old rug back on the front passenger seat, and drove on up to the village of Anston. There was a long, very steep main street, still cobbled in places, with the grey stone houses opening right onto the pavement. The cottage he was visiting was right at the far end, where the village opened up and several houses were set back behind trees and gardens. Here the street turned at a sharp right angle, but instead of going round the corner, he remembered he would need to carry straight on up a small, steep drive to the old farm cottage at the end.

He turned across the road and carefully backed into the drive. He parked and, leaving Sam in the passenger seat, apparently asleep, he got out. From here one could glimpse, past the little row of conifers at the far end of the drive, the main street of Anston snaking down the side of the fell, until it bent to follow the river along the valley floor.

The visit was a long one, to an elderly man who was cared for devotedly by his wife, herself not in the best of health; the two of them determined to remain independent and not be a burden to anyone. John wanted to help if he could.

John was still chatting when, suddenly, the wife, who

117

had gone into the kitchen to make some tea, let out a cry of horror. John, thinking she had scalded herself, leaped out of his chair. They met in the doorway.

'It's your dog. He's driving off down the road! Quick!'

John wrenched open the front door, his stomach taking a couple of switchbacks. He was just in time to see the Renault estate moving off in a slow and dignified manner down the sloping drive. And there in the driving seat, sitting bolt upright on his haunches, paws planted firmly on the steering wheel, eyes fixed on the road in front, was Sam, apparently enjoying himself enormously.

John began to run, faster then he had ever run in his life. But by now Sam had gathered speed and, as John watched in horror, he bowled merrily out of the drive and into the street.

Anston was a quiet village. There was mercifully no other traffic about, but as the car ran faster and faster down the hill, the hunched shape of Sam still steadfastly at the wheel, people suddenly appeared from nowhere, some with teacups in their hands and half-eaten sandwiches, pointing incredulously and beginning to run behind the car. It was like a carnival parade. John, running on desperately behind, black spots in front of his eyes, his chest on fire, prayed with his last vestiges of energy that Sam would not twitch the wheel, that the car would not mount the narrow pavement, that at the very least it would maintain its present majestic progress bumping and bouncing over the cobbles, and stay on the road.

Sam was now some fifty yards in front, pursued by most of the able-bodied villagers, who had not had such excitement in years; maddeningly going just too fast to catch. John could see him sitting, rigid with concentration, keeping the car in a dead straight line. He put on a spurt to try and overtake the crowd which was building up behind the car, and as he ran through them a small boy waved at him from the pavement.

'Is that your car, mister? Hope your dog's passed his test!'

It was a very long main street, running right to the foot of the fell. Ancient Britons had probably fought over it with the Romans. Now equally ancient Cumbrians cursed the street as they struggled up it every day from the baker's shop at the bottom. Sam had conquered it in a few short minutes, although at the time, for John, it felt like an eternity. But all streets come to an end. Suddenly, with a sick heart, John remembered the sharp bend at the bottom.

The baker's, where five generations of tradesmen had lived over the shop for almost two hundred years, had a small garden thoughtfully situated just at the foot of the hill. Even more thoughtful had been the provision of a sturdy picket fence instead of the usual stone wall. A stone wall would have been very nasty indeed. As it was, Sam bowled straight through the fence. Conveniently, it slowed him down enough so that he was able to park prettily in a tastefully colour-coded tulip bed which exactly matched the colour of the car.

An old man appeared in front of the car, a spade in his hand. With the phlegmatic attitude to life which characterises the trueborn Cumbrian, he regarded Sam with mild curiosity. Sam stared back, his paws still firmly on the wheel. John had reached the bottom of the hill, completely out of breath, and only yards ahead of the village's oldest resident who had put on a miraculous burst of speed, hoping to see a spectacular ending to the whole episode.

'Is that tha car, lad?' The gardener asked John with polite puzzlement.

John, scarlet with embarrassment, nodded.

'Tha want to teach tha dog te turn reet, doesn't tha?'

'Yes, I really am so sorry. I'll pay for a new fence and for the damage to your garden. I'm afraid it was the handbrake. It has a spring on it . . . He must have released it with his paw . . .'

The old man was staring in through the windscreen at

Sam who, still rigid at the wheel, looked back at him calmly, seemingly quite unmoved. John was suddenly reminded of an illustration from *The Wind in the Willows* of Mr Toad caught out by a policeman after stealing a car.

'Does tha know . . .' the old man leaned on his spade and spoke directly to Sam, who looked back at him with interest (for a wild moment John expected him to wind down his window so that he could hear better). 'Does tha know, tha's the fourth bugger this year to coom through ma fence? Though I have to say, the other three didna meak me laugh half so much so tha has done!' And his shoulders shook with silent mirth.

They drove home, not speaking. Sam had obviously enjoyed his short excursion, but John was worrying. There had been a new-looking bicycle leaning against the curb which had been knocked over as the car made its slow, inexorable progress down the village street. It had not seemed to be damaged, but John had felt obliged to leave his name and telephone number in case. And, despite the protests of the master baker, who had been digging his garden after finishing for the day, he was determined to go back with some wood and mend the picket fence, and replant the tulips. It was the least he could do, even though the baker had insisted that he had not had such a good laugh in years, and had invited him into the house to tell the tale.

They drove up the road to the village. A friend of John's was walking his dog down to the pub for an early drink. He flagged John down and asked him if he felt like a jar or two. John poured out the experiences of the day, while Sam sat smugly in the passenger seat, looking as though butter would not melt in his mouth. His friend laughed himself silly. John declined the drink and drove up to the house, seething with Sam. It was hard to see the funny side.

As he was getting ready for work the next morning, the phone rang. The voice at the other end sounded almost incoherent with rage. Closer inspection of the bicycle, the voice spluttered, had shown it to be an almost total write-off. It was a brand-new machine, the voice added hysterically. No more than a few days old. Of great sentimental value. Irreplaceable, in fact. John struggled to reconcile those apparently irreconcilable characteristics, and poured apologies down the phone. He adopted his most ingratiating tone. Of course he would pay for the damage. No, he did not normally allow his dog to drive the car. It had all been a terrible accident. He was most dreadfully sorry for the whole thing. Sam, asleep at his feet, snored loudly as though disclaiming all responsibility. He was being left at home that day, in disgrace.

The voice went on, cataloguing the damage. John adopted an even more conciliatory tone, and held the receiver away from his ear. Suddenly he heard:

'And while you're at it, young man, I've a good mind to get police to bring a charge of careless driving against your dog, on account of the fact he can't turn right at bottom of fell.'

There was a burst of laughter at the other end of the phone, and it went dead. John spent a couple of seconds staring at the wallpaper by the telephone, trying to remember something. Where had he heard that voice before? Surely, it could not be anyone he knew . . . only the people in Anston village knew anything about yesterday's incident, thank goodness, and Tina, who was in the kitchen, dressing Matthew for school. Then he remembered his conversation of the evening before, in the village on his way home. Light began to dawn. Simultaneously, he found he was staring at his son's first playschool effort, pinned to the wall by the telephone. It was a calendar made out of an old Christmas card. Today's date was the first one not obliterated by Matthew, whose job it was to cross each day off before he went to bed.

Today was April 1st.

John is still planning his revenge.

12

Sam was now the only Search Dog in the Langenhow team, and although he usually worked with Bess or other full Search Dogs on the SARDA call-outs, John had already proved worthy enough to be given sections of the fell during searches to work alone, leaving a much smaller area than would otherwise be the case to be covered by the sweep search. Throughout that year, during the long days and nights on the fell, often cut off by the fold of the hill from the rest of the team, they worked steadily together. After that first find, with Bess, it did not happen again. It was sometimes hard to keep Sam's spirits high. John knew how valuable their contribution was, but Sam, perhaps, needed to 'find', to get that 'buzz' of achievement. To quarter the fell, time after time, to find nothing, to report back emptiness, was hard. John had to make sure they had plenty of exercises with Derek or Bob Irons at weekends— when there was time; and there were the training weekends, when the dogs always found, reinforcing those crucial lessons.

Night after night . . . in one week, five nights—then work in the morning, while Sam snored his exhaustion under the baby's high chair. (Pippa, their new little daughter, born in September, had been adopted once again by Sam with fatherly delight.) John was allowed a certain number of hours away from work for rescue operations, but they were very few, and on many days he had to drag himself out of bed after two hours of exhausted sleep. It was difficult.

Sometimes, after a bad night—after a death or when a

piece of stupidity had risked dogs and men out on the fell yet again—he wondered whether he was doing the right thing. Then the team could pluck a live casualty off the fell in the teeth of savage weather, someone could live because of their efforts, and suddenly tiredness would become an irrelevance. And the joy of watching Sam, his enthusiasm unabated, quartering the fell for hours in the wet and cold, bounding up the crag, eyes alight with enjoyment, was enough.

December was the worst month. Night after night, day after day, the phone seemed to ring constantly . . . when they had no sooner sat down to tea . . . or had only just got to bed, worn out after a broken night with the baby . . . It was the sort of weather to tempt people into the fells. Brilliant, hard, sunny days; then bitter, savage nights when the temperatures above a few hundred feet could kill in an hour or two, and the frost lingered on the ground during the day, lethal, until at the first, false step . . . There were rock falls—one young but experienced climber was killed when the rocks above him, expanded and contracted by sun and frost beyond endurance, shattered over him. Despite his safety helmet, there was nothing they could do to save him. Only his two companions, critically injured, one still in Intensive Care after four weeks, would perhaps make it in the end.

As Christmas approached, things began to ease off a little. Matthew came home from his first term at school with another smeary, gluey calendar for his mum. He could not wait until Christmas to give it to her. She hung it up in the kitchen. They decorated the house with wobbly paper chains, stuck together by Anna and Matthew, and separated again (if they did not keep a sharp eye on her) by Pippa. Sam had a chocolate bone hanging on the tree. He seemed to know it was somewhere about and had twice dug the earth out of the bucket to try and find it, being severely scolded for his pains.

They bought two bottles of sherry and asked Mrs Armstrong from next door and the neighbours across the road for a drink on Christmas Eve. Another Christmas in the house . . . Perhaps when they had been in Garth How for thirty years they would stop being 'off-comers' The beginnings were there.

They raised their glasses to each other, Tina flushed from bathing the children and from the rush of getting ready in time. The phone rang. John went to answer it.

'Bill here.' The strong dialect boomed at him down the phone. 'Don't tell me it's Christmas Eve, lad. We've had a call-out. An old boy lost out on fell. He were due back at six. If you look out of window you'll see it's a reet clarty neet. Best get over here. We're meeting oop at HQ. Can tha' phone Dennis Skinner oop at Crag End?' John promised to do so. He went back into the sitting-room. The fire was blazing in the grate, Sam fast asleep in front of it. Mrs Armstrong, flushed after a couple of glasses of sherry, was holding forth to Tina about bringing up 'bairns'. She was having the time of her life.

Tina looked across at John. 'Don't tell me,' she said resignedly.

'I'm sorry, love. I'll be back as soon as I can. I'm sorry, everyone. It's an emergency. An elderly man lost on the fell—in this weather! Sam!'

Sam, deep in rabbit dreams, opened an eye and shut it again.

'Work, Sam!' That was different. He was up! Tail wagging, instantly alert. Everyone laughed.

'Your stuff's in the kitchen, John. I even left the flask out,' Tina said. 'You see, I'm a born pessimist!'

By the time John arrived at HQ it was sleeting, with the occasional burst of hail thrown in for good measure. It was not so much bitterly cold as raw, getting into your bones.

They got the full story from the police sergeant. Jed

Parkinson, the churchwarden in Kirk Langton, had set off up the fell early on Christmas Eve afternoon, to visit a housebound parishioner high up on Kittle Crag, with a present of a turkey and a plum pudding from the Welcome Club. It was a good half hour's walk up the fell, even for a man as fit as Jed who, as a retired shepherd, knew every beck and gully from Kittle Crag to Kirk Langton. There was no reason to worry even if the 'clag' was threatening to come down. It was a job he did every year without fail. Jed and old Lucy would have a glass or two of parsnip wine, to warm the cockles against the chill, and then he would make his way back down in time for Evensong at 6.30 p.m.

It was the vicar who raised the alarm, with the non-appearance of his churchwarden. The police immediately called out the Langenhow team who were the nearest, and the sergeant had himself walked up the track to April Cottage, to find that Jed had left at dusk with a basket of mince pies for himself and the vicar.

They reassured Lucy, but everyone else was very worried. Beyond Lucy's door, the track diverged, both paths running away down the fell; the right-hand one fetching up safely enough back at Kirk Langton, the left plunging after a hundred yards over a sheer crag at a place called The Rowans, into a deep ravine below.

There was little they could do while it was dark. The ravine at the base of the crag was heavily wooded, a tangle of vegetation without footpaths. If Jed had fallen there, his body could lie for months, undiscovered. But they had to try . . .

Most of the Langenhow team, torn good-humouredly from fireside and stocking-filling and Christmas Eve parties, had nonetheless managed to be on the spot at HQ within twenty minutes. Geoff and Bill between them had already worked out a plan of action. Six of the team would sweep search up the track to the cottage and beyond. Sam and John would go as far as possible into the trackless tangle at the base of the crag, two hundred feet below the divergence of the paths.

The weather had worsened by nine p.m. and, as Sam bounded up the lower path which hugged the base of Kittle Crag, funnels of wind and sleet caught them at intervals in the gaps between the trees. It was a thick night, no moon or stars, visibility almost nil. John tried to visualise the sides of the valley from the map they had studied at HQ. At one time there had been a steep path leading up over Kittle Crag at the spot known as The Rowans. But rockfall and soil erosion had steepened it to such an extent that no one ever walked there any more, and at the top, close to the head of the path, there was a warning sign. But at dusk, after a glass or two of parsnip wine, it would be easy enough to miss . . .

The wind was funnelling straight down between the crag and the far side of Langenhow Fell, which climbed away on the south side. There was a roaring from the trees. It sounded like the crowd at a football match. John sent Sam out on to the left flank, skirting the edge of the trees. If Jed Parkinson had fallen from the top of The Rowans, he would be there.

There was a stream running through the bottom of the valley. It had overflowed its banks with the rains and spread across the valley floor. John splashed through it, the going made harder by wet boots. Sam's light, like some will-o'-the-wisp, flitted about ahead. Once he came back, splashing through the brackish pools of water, and, bounding past John, went down to the fast-flowing stream to drink. John sent him ahead again, praising him. It would be a miracle if they found anything in this. He shone the torch ahead. Rain slanted back at him, and dark. Nothing else. It would not be possible to search the heart of the thicket until morning, although there was a very good chance Sam would have picked up any scent from the wood. But he made his way along the valley side, sending

126

Sam in as far as he could. John was directly under The Rowans now. There was bracken and slippery moss. He sent Sam in again and again, close up against the crag and then away in a fan shape, until he was sure. There was nothing there.

It was tricky going, but they would go on up to the head of the valley if they could. The wind had increased, sending the rain in horizontal sheets which were like a slap in the face. The valley narrowed here, and there was nowhere to walk except right by the flooded stream. Sam, facing into the wind, would pick up any scents on either side of the valley, but they would have to cross the stream to the other side—just to be sure. Sam had already plunged into the stream once. He came back, snuffling and shaking himself violently all over John, who was too wet to notice. Labradors were good in this sort of weather, and their dense, otterlike undercoats protected them from the icy water.

'I think we'll work our way back down the other side, Sam. There's nothing here. And we'll have to leave that bit of the wood till morning—just to be sure.' Sam shook himself again.

'OK, Sam.' He felt his way gingerly over the boulders at the edge of the stream. They were half under water already and even with a torch it was tricky. The stream came up to his thighs, even crossing in the shallowest part above a small waterfall. 'You go ahead, Sam. Away, find!' He gasped with the cold, feeling it even through waterproof trousers. His boots were saturated. He sent Sam farther away to the south side of the valley, where the lower slopes of Langenhow Fell came down to the stream.

It was smoother here close under the fell, and the going was easier. John started to climb higher so that he could work Sam in sweeps behind him, into the wind. The radio crackled.

'Search Dog John from Control. Report your position. Over.'

John pressed the switch on the microphone.

'Control from Dog John. I have crossed the stream at the head of the valley and am working from the Langenhow Fell side just to double-check. Have covered area under The Rowans. Nothing so far.'

The radio returned to its friendly 'squelch'. Sam came back and John sent him down the slope at an angle, to cover the rocky outcrop he remembered as being here on the south side. He shone the torch on his watch. Three a.m. It did not seem possible. He was suddenly aware of being very tired. He stopped and took off his rucksack, rummaging for the flask and some chocolate. He would give Sam something when he came back. He shouldered the rucksack, and, with the wind clawing at his back, plodded on.

All at once he was on the edge of the wood. It straddled the stream here and spilled up the fellside—a wilderness of fallen trees and thick brambles. He shone his torch over it. There did not seem to be any sort of path. The thicket washed up the side of the hill, colonising the narrow ravine. He would have to cross the stream and make his way round on the north side. He called Sam and, shining his torch along the edge of the wood, worked his way down towards the stream. It was very deep here, before it plunged into the black mass of the trees. The small streams from the fellside, themselves swollen by rain and melting snows, had fed it for days now, until here, in a deep cleft in the valley floor, it fell in a series of torrents roaring and rushing through the dark. There was less flooding here, with the steep banks, but it was going to be tricky to get across. He debated whether to walk back up the valley to where the stream was shallower—but they were both tired and very wet. The thought of fighting back through that battering wind was too much. After all, it was only a stream.

He stepped gingerly down the bank where the water seemed calmer. It was very deep here, even at the edge. His boot felt for a stone. He leaned forward. The stone turned. He felt himself pitch downward into the water.

All at once Sam was there. He grabbed his collar, and managed to haul himself half out onto the wet slab of rock. Something was amiss with his ankle. He pressed the button on his radio. There was nothing.

It was going to be a long walk back.

* * *

The landlord of the Dog and Compasses opened the pub for them. By now there were five teams assembled. He lit a fire. Maps were spread on the tables. It was Christmas morning. The Langenhow team had not slept all night. Now John was missing and they could not raise him on the radio. The team leader was getting worried. At daybreak the dogs would go out again. In the meantime, the landlord, an ex-member of Lakewater MRT, got things organised. Hot food on the house. Hot toddies. What a night!

The door opened and John appeared, looking like a drowned rat. He was limping badly. Sam was soaked. Someone appeared with dry clothes and a towel for John. Sam, too, was given a rub down and a plate of pie and mash which he gobbled in seconds, looking round hopefully for more. Then the team doctor pinned a reluctant John onto a bench in front of the fire, complaining good-humouredly the while about having to leave half his pie for someone else to pinch. He prodded about and John gritted his teeth. Eventually, he diagnosed a bad sprain, and strapped John's ankle up in a couple of yards of bandages. He was to be evacuated from the fell. John seethed to himself, but he knew that was only sensible decision. He would only be a liability.

By now the Search Panel had called in every available team in the Lake District, and SARDA had eight dogs at the ready, although now without Sam. Stuffing the turkey would have to wait. Not much chance of finding the old boy

129

alive after a night like this . . . but they had to go on while there was still that faint ray of hope.

Prayers had been said at Kirk Langton, during the Midnight Service. Jed had 'only half his insides', as he used to say— a relic of the First World War. It was a dispiriting thought.

The weather had worsened. Snow flung itself against the windows of the pub. More Land-Rovers arrived. The dog handlers, briefed on their search areas, were willing and eager to go. John felt out of it, sitting waiting for a lift down the fell, not able to take part, Sam sulking under a chair.

The latch of the snug door lifted. Everyone stopped talking. It was six o'clock on Christmas morning and no one had heard any other vehicles arrive. The dogs pricked up their ears.

A small man in a flat cap and a very wet mackintosh came hesitantly into the bar. The landlord stopped in the act of polishing a glass and stared as though he had seen a ghost. The old man looked around nervously. Then he lifted onto the bar a basket of what appeared to be very soggy mince pies mixed liberally with heather and bracken. He sighed.

'Ay, it's a clarty neet we've had. That's true. I hope tha'll forgive me, Jack. I've been wandering aboot fell all by messen, a bit lost like.' He looked around the bar. 'Looks like these poor lads have been lost an all. I must say I don't think I can meak it back to me own fireside wi'out summat to slake me thirst. I'm that wore out, I could fall down!'

13

The Annual Course in the new year was again at the beginning of February. John had just a few weeks to get himself fit. There was a lot of training to catch up on. So much depended on it. It might only have been a sprained ankle, but not only did it hurt like hell, it was a fortnight before he could get up into the fells again. So much precious time lost. And when he did get up there with Sam to start training again, he felt like an old man. His ribs ached, his ankle throbbed, and the foul cold he had managed to acquire after his ducking refused to go away. It was not an encouraging start.

For it was this year that Sam was going for full Search Dog status. He was more than ready for it. But would John let him down? Tina had been such a help. She had exercised Sam those first few days, struggling with a toddler and a pushchair, until John could get back on his feet again, and, when she could, she had carried on with the basic training . . . but it was very difficult. John and Sam were a team. Their strength lay in each other. Nothing else would really do.

In the last two weeks before the Course, however, after work was finished for the day, John and Sam would go out onto the fells behind Keswick. Volunteer bodies were hard to find, among colleagues who had homes to go to. You had to be a pretty good friend to lie out on a fell for even half an hour in sub-zero temperatures waiting to be found by a dog.

But John was lucky. Gareth Jenkins had just moved up

from Wales to Cumbria and had taken a post in the Child Welfare department next door. His dog Mandy, a huge liver-coloured Labrador, was also just coming up for her first grading as a full Search Dog. They both needed bodies for training. They took it in turns.

But John knew things were not going right. In the last week, there were two call-outs in the heavy snow. John was conscious of his ankle still throbbing as, hour after hour, they ploughed across the fell. Sam did what he was told, but the edge seemed to have been blunted. He would come up to John, excited, expectant, waiting for instructions, eyes red in the torchlight, but John no longer seemed to have it in him to respond. Sam would look at him reproachfully, John would give the routine commands, and Sam would go away dutifully and do what was required of him. He would never let his master down. But as he made his way across the fell, tail no longer wagging, his whole demeanour seemed to say, 'It used to be so exciting. I used to do it for fun. What has happened to us both to take all that away?'

As the Course approached, it was Sam himself who at last went on strike. Perhaps he was fed up. He just seemed to stop trying. In that last week of training, he did his job, found Gareth, came back to John, but without even the minimum enthusiasm. 'OK. Big deal,' he seemed to be saying. 'Who wants to be a Search Dog anyway?' John knew why it was happening. It was because he himself could not generate that excitement, that sense of achievement, that Sam needed. They were mutually telepathic. Sam knew what was going on. He did not want to try any more.

The last straw came two days before the Annual Course. John had managed to get the afternoon off on Wednesday, and Gareth had done the same. It was marvellous to train in the daytime instead of black night for a change. John felt he would do better today. He felt it in his bones. They started off up the track. His ankle hardly ached any more. He began to feel more optimistic. If only he could shake off this cold, this sense of lethargy, he would be able to do Sam justice. Gareth was ahead of them. There was a stile at the bottom of a nature trail which led off to the right. John swung up over it, feeling his bad ankle jarring momentarily. He plunged back into the snow which here, by the small stream, was deep and soft. Gareth was well ahead now. He had left Mandy in the car. It was his turn to body first. They would get a little further up the track, and then stop to allow Gareth time to take cover.

He looked behind him. Sam was not following. He was nowhere to be seen. Swearing to himself, he made his way back to the bend in the path.

Sam was sitting patiently on the far side of the stile, his tail swishing its own slow windscreen wiper track across the surface of the snow. John whistled to him. He did not move.

'Come on, you daft bugger. We're waiting for you.' Sam looked at him steadily, but did not shift. His tail had stopped wagging. John called him again. Still Sam sat there in the snow, looking back at him. There followed a very unpleasant few minutes during which John called Sam every name under the sun. Their relationship deteriorated to an all-time low. Sam, his nose a few inches from the stile step, steadfastly refused to cross it. He was too bulky to squeeze underneath and, if he would not jump it by himself, there was only one other alternative.

John looked over his shoulder uneasily. This was too much. Gareth was out of sight, hopefully gone to ground. The fell was apparently empty. He would never live this down!

He crossed the stile and, bending at the knees, gathered Sam in his arms. Sam licked his face and buried his nose in his ear. John climbed back over the stile, seething. Sam, thoroughly enjoying himself by now, laid his head on

133

John's shoulder and closed his eyes with a deep sigh.

'I will never forgive you for this,' John ground out, through clenched teeth. 'You weigh a ton. Don't you ever do this to me again.' He felt a tail thump against his arm. 'I suppose you think it's time you complained. Well, I get the message. Just don't push your luck.' He bent down with a groan and put Sam down in the snow on the far side of the stile. 'There! Will that do you? Now, shall we start again?'

They understood each other. The last two weeks they had been like two cogs in a millwheel, which were; no longer synchronised. Jarring the whole machine. Breaking up. Now, suddenly as John sent Sam away up over the frozen fell, it was back: that magic, indefinable thing. You could see it in the way Sam moved, the way his nose lifted, savouring the wind. John was no longer conscious of his ankle throbbing, of his streaming cold. This was what it was all for. If only they could keep it up!

*　　*　　*

The four days of the Annual Grading Course were to be held once more on the slopes of Helvellyn and the fells around Thirlmere. It started raining on Friday, and went on, steadily and sadly, for four days. The dogs' coats were dark with rain. The assessors stood about in hunched groups, licks of hair plastered to damp foreheads, looking as though they wished themselves anywhere but out here on the slopes of the fell. They were going to be a hard bunch to impress.

Gareth and Mandy were allotted the area of fell adjacent to John and Sam. It was good to have a friend near. Gareth was even more nervous than John. It was an unfamiliar area for him, and he must have felt much as John had felt on his first Novice weekend, except that Mandy was rather less likely to do anything outrageous. She was a good, steady dog, but Gareth was worried sick. He had recently lost his wife and, having no family, the dog had assumed enormous importance in his life. As they stood together at the base of the fell, waiting to set out on their first search, John could feel the nerves radiating from him. It was certainly affecting Mandy, and even Sam was getting jumpy.

John tried to calm his own nerves. It would not be like last time. Sam was now a good, mature dog with real-life search experience. He could do it, if only John did not get himself into a state. It was just no good worrying, thinking about all those years of training, and about how little he had been able to fit in during the last few weeks. His stomach contracted. He stood watching the first dog, a small spot on the scree, appearing and disappearing through wisps of claggy cloud, her owner fighting her way across the lethally slippery rock towards her. What a pig of a day. Rain was running down his neck. His cold had started up again. Would they ever make it?

Mandy was to go first. John was relieved. It would give them a few minutes to get themselves together. The rain had begun to turn to freezing hail. He watched Gareth make his way up the steep path beside the beck. He would have quite a large area to cover in a limited time. John felt sorry for the body, lying out there in this . . . even in a bivvy sack it would be pretty wet.

He lost sight of Gareth. The next moment it was his turn. The body was already in place. One of the assessors showed him on the map the area he had to cover. It would be on the far side of the fell, going down into the fold in the hill. He began to walk up the path, the rain beating on his face. Halfway up, the cloud cleared enough for him to catch a glimpse of Mandy away over to the left. She seemed to be having difficulty on the slippery crag. It was awful ground to walk over, especially when it was wet. And Gareth had done so much of his training in the softer hills of South Wales. It was going to be hard for him.

Then he forgot about everything except concentrating on what was going on. The rain was freezing as it hit the rock, half-hidden as it was under bracken. You could turn your ankle without even thinking. And the ground underneath, loosened by rain, was lethal, too. Sam was going carefully, head up into the driving rain, concentrating, too. Sam never let you down. He knew when it was important. John stopped worrying. They could only do their best.

They had got their old rhythm back. Was it just his imagination that Sam had staged a protest, that day on the nature trail when he had refused to cross the stile? Or had Sam really known, in that strange way of his, that John needed to be reminded? Whatever the true explanation, John's lethargy had evaporated, almost magically. Dog psychology? He was almost sure he did not believe in it.

Sam was standing in front of him, looking at him, telling him they had drawn a blank. They would have to go higher and work across the top of the crag into the wind. John lifted his arm.

'Away, Sam. Away, find.' But Sam was already scrabbling upwards towards the skyline, tail going crazy. He had figured it out for himself.

They worked for almost an hour across the top of the lower fell. Occasionally, the clag cleared and he could see into the valley below, and could just pick out the orange hood of one of the assessors standing miserably, no doubt, in the driving rain. But for the most part, they both forgot. This was just another search.

John's legs were beginning to ache with the constant tensing of muscles, clambering over the wet scree. But every time, Sam bounded ahead, seeming as fresh as when he began. John pulled back his glove and peered at the rain-spattered face of his watch. Only another half-hour till their time was up! They had better get a move on. But it was no good getting worried, cutting corners. That was what the assessors would be looking for, if indeed they could see a blind thing down there in the valley.

Suddenly, Sam was there, springing out of the smoking mist like a very solid genie. He was chomping his jaws, wild with excitement and pride.

'Speak, Sam! Speak!'

Sam chomped again.

No doubt about it.

'Show me, Sam. Good dog.'

Sam turned and scrambled across the wet fell, making for a granite outcrop which reared over the valley. John followed as fast as he dared, his feet slipping dangerously on the treacherous ground.

'Show me, Sam! Show me!'

There was a cave under the rock. A flash of blue. Sam, standing by, radiated satisfaction, his jaws chomping with that peculiar bark which meant only one thing: 'I've done it!'

No one said anything: they rarely did. But John knew they must have done OK. The conditions had been pretty poor, worse than most. All they had to do was keep it up.

Gareth was not so happy. Mandy had tried her best, but she had found the ground hard to adjust to, and had stuck too closely to him. Gareth had tried to send her away, but she had been uncertain, troubled.

'Perhaps it was me,' he said gloomily, as they towelled the dogs down, and drank scalding tea out of John's thermos. 'I was pretty nervous. I don't know. Every time I got cross, she just stuck tighter to me. We found all right, but I know she wasn't working as she should.'

John nodded.

'The more worried you get, the worse they are. I know, it's the same with Sam. You just have to try and relax. Pretend it doesn't matter.'

'But it *is* so important. We all know that,' Gareth said gloomily. 'Years of training. I don't know if I could face another year as a Novice. I think we'd both go off the boil.'

'I know what you mean,' John said. He could hardly bear to think about it. All that training. Could he go on if they failed? He did not know.

The next two days passed in a daze. They always seemed to be cold and wet. There was snow and occasionally gleams of sunshine piercing the gloom. John fell into bed at night, after a pint or two in the bar, dreaming searches almost before his head touched the pillow. Over and over again, he would send Sam away for imaginary bodies which were never there. He saw the faces of the assessors. They would shake their heads. Sometimes in his dream he would lose Sam in the clouds, and have to set off across an endless fellside, calling and calling. Then the alarm would go off and he would wake, feeling worn out, hearing the rain on the window, other doors opening and closing, until a huge breakfast and coffee miraculously restored him to life.

So far, actually, they were doing well. Despite those nightmares when the assessors stared at him, shaking their heads, Sam had found well within the allotted time. On the second day they had been given three searches. The same on the third day. Now there was just the Tuesday morning to get through.

The weather had changed overnight. It was bitterly cold. The sheets of water in the car park down from the hotel had frozen into a lethal covering of ice. Snow blew in the wind. They all put on extra layers of clothing.

The fells were transformed. Overnight the bracken had frozen frost-brittle, the grass rimed with white. And as the cars arrived, the first snow flurries speckled the wind-screens. John opened the door, and a blast of icy air hit him.

Gareth came over to the car as he was lacing his boots. John could see he was worried. Things had not been going too well. You could not put your finger on it. Mandy had found every time so far, but she was still clinging to her

138

master, unsure, tense. On a search over a wider area, how would she behave?

The assessors seemed to have that in mind. They had chosen large sections of fell, with a number of close contours on the map. This would mean the dogs would be out of sight of the owners for long periods of time. It would put a lot of responsibility on the dogs themselves.

They split up and began to make their separate ways up the fell. Already the rain-sodden picture of the day before had changed as the first dusting of white draped a cloak over the hills. It was extraordinary how suddenly the fell was different. Lonelier, harder, more dramatic. But infinitely beautiful. John loved to see the way the fells showed their bare bones . . . suddenly no colour but grey granite where wind shaved the snow cleanly as soon as it lay, yet on the bracken it stayed deep, unspoiled white.

The skeleton of the fell showing through; that was it. But, practically, he could have done without the faceful of snow which filled his eyes and mouth, causing him to lose sight of Sam momentarily, and the hidden faces of glassy puddles ready to slip him up. It was always the same, the wilder, the more achingly beautiful, the more treacherous this country became.

He sent Sam away to the right, into the wind, which screamed down from the tops. It was getting up. It was amazing, too, how in half an hour those drifts were forming in the hollows by the beck, smooth as icing sugar, their tops waved by the wind. Sam plunged in, spoiling their perfection with scrabbled paw prints, snuffling as the powdery snow got up his nose. Then he was on the ridge, snorting and shaking his head, 'feeling' the wind from the crags.

John wondered momentarily how Mandy was faring. Then he sent Sam away, over the fold in the hill, down into the far valley, the map in his mind. There was a dry stone wall beyond it, straddling the fell, keeping the sheep in the intake fields while the weather was bad. Many would have their lambs already, because here, in this hard country, the Herdwicks dropped their lambs from Christmas onwards, in the harshest months of the year.

It might be that the body would be beyond the wall, but that would be no problem for Sam. Astonishingly, a good dog could scent a body on the other side of a stone wall with ease. It helped up here in this country, where the ancient walls were everywhere, keeping alive the old boundaries which the Celts had made.

They worked on steadily. Most of the time Sam was out of sight. He was used to this country, and confident enough to work by himself, only returning to John occasionally for more instructions. The wind increased still further as they came out onto the exposed face of the fell. Here the snow was already lying several centimetres deep. Suddenly Sam was by his side, 'indicating', but in a puzzled way, as though he was saying, 'I've picked up something but I'm pretty sure it's not what you're expecting.'

John put his hand on Sam's collar.

'Show me, Sam. Show me.' He shouted against the wind. But Sam did not immediately bound away as he would normally have done. He lifted his nose and barked. Most unusual behaviour. Then, over the ridge, to John's amazement, two small figures appeared, slipping and sliding down the fell. Others followed them until there was a crocodile of children all struggling across the whiteness. Some were crying. He could hear that even above the wind. Himself slipping on the icy slope, he followed Sam who, having delivered his message, was racing towards the children through the snow. John realised he must have picked up the scents of the whole party on the move. They were directly upwind. But what on earth were they doing out here, in such conditions?

As he came closer, he saw how young they were . . . junior school children, he would have guessed; and clad only in the flimsiest of anoraks and trainers, and tight jeans —no use at all in this intense cold. And there were two adults with them. They all caught sight of him at once, and the children sat down in the snow, where they were, obviously exhausted, while Sam, who adored children, bounded round them in circles. One child John saw had got himself stuck on an icy bit of the ridge, where it was very steep. The man in the party was trying to coax him down, but he had had enough and was clinging to the frozen bracken at the top, sobbing loudly.

He looked up as John appeared through the driving snow.

'Oh, thank goodness! We were having a bit of a time getting down. I could do with a hand. We only came up for a nature ramble. Didn't expect this!'

John swallowed down his anger—a nature walk! No wonder there were so many accidents when people behaved as though the mountains were their own back yard, and with young kids in tow as well! Between them they lifted the crying child down the slope. The children were obviously very cold and wet, and it was still a long way down to the car park. They were up to two thousand feet even here, and the slope was steep. John got on the radio and requested help. Then he and Sam between them began to pick their way down the fell, Sam's antics doing more than anything to cheer up the frightened children. Halfway down, three other dogs and handlers, who had been waiting their turn or had finished their work, appeared below, plus an assessor (who had seen the party appear over the brow of the hill while he was sitting in a tree watching John!).

John led them to the bottom of the track by the stile and left the others to it. He was pretty sure the two teachers would not go home without some opinions being expressed about the risks they had been taking and the irresponsibility of taking children up in any conditions without proper clothing. In different circumstances that group could so

141

easily have been the subject of a full-scale search, with the likelihood of a tragedy at the end of it. How long would those children have survived, in their pathetically inadequate clothing, in a white-out on Helvellyn? That kind of thing happened all the time.

The radio crackled.

'Dog John from Control. You only have half an hour left. We do not have radio contact with your body. She has been instructed to walk off the fell at 14.30 hours.'

He increased his pace up the path, sending Sam away directly ahead of him, although they had covered this ground already. In fact, when they had come upon the school party, they had covered three-quarters of the area. Would the scent from so many 'bodies' obscure the air scent? Ahead of them, in the already fading light, John could see through the snow flurries, the churned-up whiteness where the children had come down the mountain. Sam, not even pausing, ran straight across and down into the narrow ravine on the far side. John could tell from the line of his body, from the way he moved, that he had picked up something else, and all those confusing, 'moving' scents were not fazing him at all. He pulled back his cuff and looked at his watch. Quarter of an hour to go. Not very long. They had better be on the right track. The assessors would be sympathetic, but it would still not look too wonderful if he did not find the body at all. So much depended on whether he had really done well so far or whether it was just his imagination. Even if you 'found' every time you still might not be doing it properly.

The wind was winding itself up into a full gale. Snow flurries skimmed the surface of the fell, fetching up against the face of the nearest crag, or sweeping up into his field of vision like a swarm of angry albino bees. They stung at his face, making his eyes water.

Sam was there, beside him, chomping his jaws, tail going like a pendulum.

142

'Show me, Sam. Show me!' They fought their way across the fell and, slipping and sliding, down the side of the ravine. The snow was deep already, soft seductive powder snow. You plunged down into it. It was quiet and beautiful. But in different circumstances it would be death. He felt his way over. Sam, his back legs pumping like pistons, had already reached the far side. He stood watching his master, tail wagging impatiently, barking occasionally as though to say, 'Come on, slow coach! There isn't much time.'

Sam disappeared around the far side of a granite outcrop, and reappeared almost immediately, leading John in. He followed, scrambling over the snow. Blue and orange. A girl in a bivvy sack. Pleased to see them. Just in time. John made a fuss of Sam, and felt in his pocket for the choc drops. Sam was licking the girl's face as though she were a real casualty, just as pleased to have saved her. To him it was all the same.

Sam was towelled down and given some food. John had had no chance to see Gareth. His car had gone by the time he got back to the car park. It was dark by now, the snow coming down fast. The assessors were packing up. He was the last. He swung out of the car park, and down towards the lights of Keswick.

The tea was at four o'clock. He would have to hurry if he was to be in time. The roads were already becoming tricky, and he had to concentrate on his driving. Suddenly he was terribly weary. It all seemed to have been going on for a very, very long time. They had done their best. There was nothing more either of them could do.

He settled Sam in the car in the lakeside car park, and made his way over to the hotel. As he pushed his way through the doors he saw Gareth standing by the desk. Something about the set of his shoulders warned him what was wrong. He picked his way through the jumble of boots in the entrance.

'OK, Gareth? How did it go?' Gareth turned to face him. He was a hefty six-footer who had played rugby for Wales, but John was embarrassed to see how near he was to tears.

'They had a quiet word, John. Mandy hasn't got through. I'm on my way home.'

John's heart sank.

'Gareth, I don't know what to say. I thought you were doing all right.'

Gareth shook his head.

'It was me. I know that. I was so uptight. It had come to be so important to me, I must have got so tense. It was worse today. Poor Mandy, she would hardly leave my side, and the crosser I got, the closer she stuck by me. You know how it is.' He choked suddenly. 'I don't know. I feel as though I've let her down. She's such a wonderful dog.'

John stood there, feeling awkward.

'Come and have a drink. We've got time. You'll feel better in a bit. And there's always next year.'

Gareth shook his head.

'I think I'd like to be getting back, John. Thanks all the same. I want to give Mandy a good hot feed. Say thank you, like. After what she's done for me the last few days . . .'

'But you will go on next year, won't you?' John asked. 'Mandy's such a great dog. It would be such a shame. Don't stop now.'

Gareth swung his rucksack onto his back.

'I don't know, John. Perhaps I'll feel better in the morning. I want to think about it for a while. Perhaps it's me who's not cut out for it . . . who knows?'

John stood in the hall watching helplessly as Gareth pushed open the glass doors and disappeared into the dark. What could he say? He knew that if no one had said anything to him yet, he and Sam were probably OK. They had worked hard. They deserved it. But if it had not been for Sam's protest, half-comic, half-serious, at the bottom of the fell, it might well have been he himself walking out

144

into the snow, to face the long drive home, knowing they had failed.

The aroma of steak and kidney pie came wafting down the corridor from the dining-room. He would have to hurry, or he would miss the tea. Suddenly he realised he was starving!

The roll call of dogs who were to be awarded full Search Dog status was very short. This year only three dogs had got through. Sam's name was at the head of the list.

John went up to receive Sam's badge and certificate. Suddenly he felt very tired. It had been a long, long run. Tomorrow they would celebrate: Sam would have a giant chew-bone and a new collar. Tomorrow he would be ecstatically happy. But for today, thank goodness it was all over. They had made it at last.

14

It was very different now. Although Sam had been 'solo' on a number of limited searches, he and John were now given much wider responsibilities. They were completely on their own with no back-up except the radio, and expected to cover a vast area of fell. In the first two months up to Easter, Sam was involved in seven searches, four with a major SARDA commitment, the other three assisting the Langenhow team in a search capacity. And there were eight other call-outs for John, which were rescues without a major search element.

The weather had worsened, and day after day of torrential rain produced its own problems—rock falls, floods, sodden slopes where even a properly equipped walker might slip and fall. But the truth was that many of the people who were tempted to 'go for a walk' on the fell paths did so in hopelessly inadequate footwear—town shoes, plimsolls, even sandals. Four of the rescues involved bringing visitors down the cliff face, who had broken legs, arms, or ribs from slipping in *smooth-soled shoes*. Those words must be graven on the heart of every MRT member! One girl slipped from the path in heeled shoes and fell thirty feet into a rock pool below. She was very lucky to survive with 'only' a fractured skull.

Sam's first outing as a fully-fledged Search Dog came at the end of a hectic week. There had been severe flooding in the next valley and, when the waters had receded, a white mini-van was found parked at the bottom of the fell. It was identified as belonging to a Manchester man who had been reported missing from his home a week previously. He had been suffering from severe depression.

After an initial, very wet, search, during which the team leader reported that if he had to go on wading through water up to his knees, he would start wearing flippers instead of walking boots, all the available SARDA team members were called in. Sam was assigned the north side of his own valley but he and John drew a blank. The search went on all day, and into the next. The following afternoon Search Dog Ben, a Collie, and his handler, came upon two medicine bottles and a scattering of small red capsules, half-dissolved, by the side of the tarn. A few yards further on they found a coat; then a tie, one shoe and then the other. At last, huddled in the lee of the rocks, where he had fallen asleep for the last time, the body of a man.

It was a sombre stretcher party which brought him down. The rain cascaded out of a leaden sky, unrelenting. The team were much affected as always, but could not even summon up the usual morale-boosting jokes which helped sustain them through a 'foxtrot'. At the base of the fell, the wife was waiting in the police Land-Rover. They came down slowly and carefully, the dogs disturbed and uneasy, sticking close to their handlers.

The man's wife threw open the door of the Land-Rover and came running. They looked into her stricken face, but had no answer.

'Why?' She grabbed John's arm as he and the others lowered the stretcher to the ground. 'Why come all this way . . . ? Why come to all this . . . ?' She looked around at the grey fells, shrouded in mist. 'He'd never been up here in his life. Why come now? Tell me! It was suicide. I know it was. He's been so depressed . . . about money, the divorce . . .' John looked at her helplessly. The sergeant came and took her away.

That night he and Sam sat together by the fire, John saying very little. Tina, sensing his mood, brought his supper on a tray. After the children were in bed, she perched on the arm of the chair and put her arm around his shoulders.

'I know it's easy to say—but you mustn't get involved. You won't do a good job if you do . . .'

John went on staring into the fire. At last he said:

'We're so lucky, Tina. We have so much. I can't help thinking about that poor devil—dying alone . . . It *is* so lonely up there, sometimes . . . All the way from Manchester, with a couple of bottles of barbiturates in your pocket— just to kill yourself. She asked me why—but I couldn't tell her.'

'Perhaps he just wanted a little peace,' she said gently. 'He may have been happy in the end.'

'Even so, it's such a stupid waste . . . we could have saved him. I don't know.' He rubbed his forehead violently with his hand. Sam's head was on his knee. He had not slept all evening; merely sat, his eyes on John, watching.

'Sam needs a walk. It's cleared up now. It'll do you good. And, John, I know it's hard . . . but just don't get so involved!'

* * *

In that first year, too, there were moments of almost farcical relief. Like the woman who took her dog for a walk up Langenhow Fell in deep snow. They had climbed almost to the top when the dog—an Old English Sheepdog— became immobilised: its fur was balled up completely with chunks of rock-hard ice; it was totally unable to walk. Mrs Prendergast was forced to stagger down the fell, carrying some ninety pounds of snowbound dog. She wandered off the path where she was found, exhausted, by Bob Irons and Bess, fortunately before darkness fell. The police had been alerted by Mrs Prendergast's husband almost as soon as her return was overdue.

'She's a very stubborn woman. I warned her what would happen. And now I suppose she'll say I was fussing. I just hope she's learnt her lesson!'

That episode at least ended happily.

And then there was the strange business of the lights . . . which was never explained, to the satisfaction of Miss Carter, or anyone else . . .

*   *   *

It did not feel like a day which would end in disaster. But then, unless one is blessed, or cursed, with second sight, it very rarely does.

It was a beautiful spring morning. For weeks Margaret Carter had promised the choir of St Oswald's that on the Saturday before Easter she would take them walking in the fells. With her usual efficiency she had circulated lists of essential equipment, called on parents, planned everything down to the last detail.

Then, as usual with the best laid plans, something went wrong. The organist, Percy Carruthers, a keen and sprightly walker himself, suddenly went down with quite unseasonal 'flu. Miss Carter was adamant that she must have another adult in the party. She was as fit as a flea, but loose scree and a broken ankle were no respecters of persons. Miss Carter canvassed the village but there were no volunteers. It was the start of the Village League Cricket Season on that particular Saturday and most of the able-bodied fathers would be too occupied to take time off for fell walking.

Miss Carter was distraught. It seemed as though the expedition, planned for her only free Saturday for weeks, would have to be postponed indefinitely. And the children had been so looking forward to it. Steven Crowthorne had even requested a rucksack for his birthday in honour of his first expedition into the fells. Miss Carter was constantly surprised by the fact that so many fellside villages contained children who had never in their lives explored the hills around them; in fact hardly seemed to notice that they were there. After living most of her life in an industrial town before coming to teach at the church school attached to St Oswald's, there would never be a time when she took the fells for granted. But she supposed it was the same anywhere. If it was all around you, you just accepted it.

Then, after pouring out her troubles to the vicar over tea, she was amazed to find a volunteer from a most unexpected quarter. The Reverend Parkston was full of enthusiasm. In his youth he had been a great walker. It would be marvellous to be up in the fells once more. If he might accompany them . . . ? Miss Carter's eyes strayed inadvertently to the vicar's waistline where the clerical waistcoat did little to disguise the ample evidence of middle-aged spread.

'But, vicar.' Miss Carter took a deep breath and cut ruthlessly into the Rev Parkston's excited eulogy. 'How long is it since you did any serious walking? We are hoping to cover at least ten miles. Do you think . . . ?'

The vicar waved the sugar bowl dismissively, before extracting a fourth lump for his tea.

'It seems like only yesterday that dear Bunny and I walked across Striding Edge in a blizzard. What days they were . . . ! and of course I keep very fit with my bicycle round the village. Nothing like riding a bicycle to keep the old legs in working order. Eh, Miss Carter?'

So Miss Carter, with some misgivings, agreed.

But as they began the long scramble up the side of Langton Raise, it rapidly became obvious that the vicar was proving to be a bit of a liability. A middle-aged bachelor, much spoiled by his housekeeper—that's how Margaret Carter saw it. A stricter diet; a little less of the convivial sherry on his parish visits, and that slight tendency to overweight, that tinge of purple in the complexion, might be avoided. Miss Carter had lived by a stricter code and had little patience with the waverers of this world.

But she was fond of the Rev (as she called him to herself) and was glad enough of his company. If only he weren't so slow! The boys and girls were leaping ahead, while she was forced to hang back with the Rev, who pounded up the fell path wheezing like the organ of St Oswald's on a damp day.

They stopped for lunch by Ramble's Tarn, a secret rock pool hidden from view by the fells around. It was a perfect day. The sun beat down on them hotly. The vicar removed his sports jacket and lay back, much out of breath. Miss Carter occupied herself with setting up the primus stove and distributing the picnic. The children sat around on their cagoules, dipping their feet in the icy waters of the tarn and screaming, the sound ringing round the crags. Miss Carter smiled at them fondly. What dear children they were—and Steven Crowthorne's upper C♯ rising into the hammer-beamed roof of St Oswald's more than compensated for his occasional regrettable lapses of language. It was fortunate, perhaps, that the dear vicar was asleep. Miss Carter thanked her lucky stars that years of teaching in a dockside primary school in Gateshead had taught her a thing or two and left her virtually unshockable. The Rev, on the other hand, had more than a touch of other-worldliness about him.

Miss Carter got out the map and spread it on a flat rock. The children gathered round.

'I thought we would take the footpath up from here towards Shale Rock.' She traced with her forefinger. 'It's a goodish walk but as we have deviated somewhat from our planned route—' she did not like to add that the vicar's slowness of pace had rather upset her plans—'we should get back on to the path as soon as possible. I have left a route with Mrs Thomas (the Rev's housekeeper) and we are expected back at four-thirty . . .'

For several minutes, Steven Crowthorne had been trying to attract her attention. Drat the boy. She waved him away as though swatting a wasp, and continued, a trifle louder.

'As I was saying, we had better get our skates on if we're to get back on our proper route and be home by four-thirty p.m. . . .'

'Miss, miss!' Steven was not to be put off.

'What *is* it, Steven?' Miss Carter enquired with more than a touch of asperity. 'I have already explained that you have eaten positively the *last* egg sandwich . . . In fact you seem to have consumed the last four, to my certain knowledge . . .'

'Miss, miss, I think the vicar's dead!'

The pandemonium which broke out more than exceeded Steven's wildest expectations.

Miss Carter, stirred by a rare ring of sincerity in Steven's voice, leaped to her feet scattering maps and tupperware, and ran over to the recumbent figure of the Rev, pursued and surrounded by excited and awed children.

'Oh, miss, he's a horrible colour. He's blue. He is dead. He is!'

One or two of the younger ones began to cry, the ghastliness of the situation beginning to eat away at their initial excitement. There were a few more sobs; silence.

Miss Carter was unbuttoning the Rev's shirt, loosening his collar. Luckily today he had on an ordinary shirt and tie. Her fingers were like thumbs: she could not have coped with a dog collar.

She put her ear to his chest: a faint flutter.

'Lend me your mirror, Cheryl!' Dear child, always checking her eye-shadow every five minutes.

'Yes, miss. What for?'

He was still breathing, the glass misting slightly.

'Is 'e still breavin'?' Steven's anxious voice was by her ear.

'Yes, Steven, I think he's still alive. I think he may have

had a heart attack.' She was shaking now. Reaction.

'What are you going to do, miss? Give him the kiss of life?' Glorious scandalous implications had already occurred to somebody, now the immediate terror was past. Miss Carter took a micro-second off to marvel at the resilience of children.

'We had better put him, so, on his side. And get him into the sleeping bag—I saw you had it, Cheryl—and some of you lend me your cagoules. We'd better keep him as warm as we can.' She looked at her watch. It was two-thirty p.m. Another two hours before even the faintest suspicion might arise down in the village that all was not well. And she was in a ghastly dilemma. The eldest choirboy—Steven —was only thirteen. Dared she send him—with a companion—down through the fells, a journey which would take a good hour and a half, supposing the weather did not worsen, to get help? Or should they all go down, leaving the dear vicar in the tender care of Steven and a few other responsible boys? Either choice seemed appalling. She made a decision.

'Well, children. We must stay here with the dear vicar. It's a lovely day. We can put up the tent.' Miss Carter was glad for the first time that she had had the foresight to carry a two-man tent in her backpack, although its extra weight had caused her considerable discomfort on the way up. 'There are only eight of us. We can take it in turns to sit in it. It'll be the most tremendous fun.' She forced herself to keep the tremble out of her voice. Dear Cheryl was already sniffling into her hanky, and the younger ones looked as though they might begin again at any moment. She shepherded the children to the grassy area on the far side of the tarn, under Cheryl's care.

'We ought to go down and get some help for 'im!' Steven Crowthorne had crept up beside her and was staring, white-faced, at the prone body beside her.

She shook her head.

'I don't see how we can, Steven. I can't leave the vicar and I can't leave you! So we're rather stuck, aren't we?' She smiled at him, a trifle too brightly. 'Once Mrs Thomas realises something is wrong . . . I gave her *strict* instructions to contact the Rescue Service if we were even half an hour over schedule. I had allowed plenty of time, you see, Steven.'

Steven looked doubtful.

'But, miss, how are they gonna find us when we ain't on the proper route? We was supposed to go up this way. That steep bit. You said.'

Steven produced a brand new map, obviously another birthday present. Miss Carter was momentarily touched.

'I've been studying it. We should have gone up here and over to this bit here—oop near the top. But 'stead, we're here.'

'Yes, Steven, I'm afraid you're right.' Miss Carter was forced to agree. 'I had intended to get back on the original route—here—this afternoon.'

'I s'pose it was because of the old vic getting puffed out, you didn't think we could make the tops?'

'Yes, Steven, I'm afraid so.' Miss Carter was suddenly overcome by despair.

'So if them Rescue people come looking for us, they're gonna look in the wrong place?'

Miss Carter nodded dumbly. The whole thing was an awful mess. If only she had been more ruthless.

Steven leaned over the Rev and gently pulled back the cagoule from where it shadowed his face.

'Will 'e die?' he asked flatly.

'I've no idea, Steven. But he ought to have help right away. And we can't move him. We're on the horns of a dilemma, you might say.'

'Me and Cheryl. We're the eldest. We'll go down. Cheryl's been walking with her dad. She'll know the way. And I got me map. And me compass.' He fished yet another

154

birthday present out of his top pocket and displayed it proudly.

'Oh, Steven, I can't possibly let you go down. It's far too dangerous by yourselves. You might get lost. Anything might happen.'

The vicar stirred suddenly and made an awful noise in his throat. Miss Carter bent over him.

'Oh, dear! Oh, dear!'

'That's it, miss!' Steven got to his feet. 'It's us got to go down. Someone's got to help the poor old vic, and me and Cheryl's the only ones as can do it. You gotta stay here and look after the littl'uns. We only has to go down the same way we come up!'

Miss Carter was full of misgivings. She insisted on outlining the route on Steven's map, on showing him how to use the compass properly, on giving him the only polybag for the two of them to share in an emergency, if they should become lost.

They set off down the hill in bright sunlight. Miss Carter looked at her watch and was amazed to find that only an hour had passed since the vicar had collapsed, and it was still only three-thirty. How long would they have to wait, even supposing the two children got down safely? Had she done the right thing?

The vicar was lying on his side now, propped against the rocks. She dared not move him. It was just a case of keeping him warm and quiet. There was nothing more she could do. The children were much occupied—on the far side of the tarn—with trying to erect the tent. Their fear and shock had receded. If the weather held . . . if Steven and Cheryl found their way back to the main road and to the nearest telephone box (in the village main street of Maston, half a mile from the path, she had instructed Steven) . . . if they were found soon . . . All might yet be well.

Just then, the sun went out, like a light suddenly extinguished. Miss Carter began, quietly, to pray.

*   *   *

Mrs Thomas had done a nice tea. It would be a treat for the poor little blighters, half of whom never got a square meal at home as far as she could see—all take-aways and school lunches. She had boiled the kettle twice before she began to get really worried. There was no doubt about it . . . . they were overdue. Miss Carter had said four-thirty, and Miss Carter was not the sort of person to make a mistake.

Five-fifteen came. Mrs Thomas checked both the clocks, phoned the Speaking Clock, decided to boil the kettle once more for luck, by which time the sound of the Rev Parkston's triple ring at the bell would dispel all her worries . . .

Miss Carter had said, 'If we're not back by five, phone the police.' She stood by the phone, besieged by doubts. She would feel such a fool . . . They would probably walk in the moment she picked up the phone . . .

She lifted the receiver experimentally, one ear alert for the vicar's footsteps. The local policeman in the village . . . she could not bring herself to telephone the Rescue Service . . . when any moment they would walk in the door . . . but perhaps just a word with the village bobby would not do any harm.

*   *   *

Within sixty seconds of Mrs Thomas alerting the local police, it was decided that this would be a Search and Rescue operation. Police Constable Nicholson could see Langton Raise from the wistaria-covered sub-station, from which he operated three times a week, and the sight filled him with foreboding. It was shrouded in cloud, so that, for several minutes at a time, it seemed to vanish altogether, leaving an empty skyline.

He knew Miss Carter; and if Miss Carter said she would

be back at four-thirty then you could set your watch by the time she came marching down the fell path to the village . . . so if it was now five-twenty-five and she hadn't appeared, something was badly wrong.

Within five minutes, both the Langenhow and the Langton Raise teams had been alerted and were requested to bring equipment for a full-scale search. Police Constable Nicholson jumped in the police mini and sped off down the village street to interview Mrs Thomas, who was still dithering about whether she had done the right thing. She gave him the route map which Miss Carter had kindly left with her that morning. The Police Constable thanked her and said it might well be some help, but you never knew. Mrs Thomas said that at least it was still light and the weather was holding. He agreed and told her not to worry. As he drove away he looked up at the innocent pale sky above the village, where the vast bulk of Langton Raise usually dominated the view. There was nothing. The fell had disappeared.

*   *   *

Steven Crowthorne was furious with Cheryl. Just because they could not see where they were going, it did not mean he did not know where they were. She was too stupid to grasp that he knew how to use a compass; that he had memorised the route inside his head; that despite the thick mist he was quite confident of the path. She was doing her loony act, sitting on a rock smudging her mascara, having a good howl. He had a good mind to leave her—after all, there were more important things than Cheryl Atkinson with a dose of the jitters . . . but he could not. And now, with her tears and her whining, he was beginning to get a bit confused himself.

He forced himself to think, squinting down at the compass through the fog. A minute ago, the sun had

beaten down on them so hotly he had taken off his jumper.
Now they were shivering, damp and chill. He could shake
Cheryl. He shouted at her to put on her cagoule, to hold his
hand and just follow . . . at least there was still a semblance
of a path glimmering faintly in front of them. He pushed
away the thought that it was not at all like the wide, well-
beaten track they had followed up to Ramble's Tarn . . . as
long as they kept walking downwards, slowly, carefully,
they could not go wrong . . .

*   *   *

There was nothing for it. They would have to move the
Rev. Any longer in this bitter, chilly mist and he would die
of exposure! Miss Carter organised them into a little team
of lifters, a few on each side. Carefully, very carefully,
keeping the prone body as flat and still as possible, the
children and Miss Carter staggered towards the tent. He
was an unbearable weight and, try as they might, his feet
were almost trailing on the boulders by the time they got
him to the shelter of the tent walls. Miss Carter had chosen
well. The tent had been erected in the most sheltered spot:
protected from the wind, and with an overhang shielding
them from the worst of the rain which by now was driving
into the rock walls above them.

Miss Carter, for the twentieth time, looked at her watch.
It was half past six. By now, if all had gone well, Steven and
Cheryl would have alerted the Rescue Service and given
them a good idea of their position with the help of the map
she had marked so clearly for him. In a few moments, they
should hear the first sounds of a search party coming up
the fell. She explained all this to some of the weaker
brethren, who were inclined to be a trifle tearful, as she
was sharing out the Kendal mint cake. Only three little
ones at a time could squeeze into the entrance of the tent to
keep warm, so they took it in turns.

They decided to sing. It seemed an appropriate thing to do, and practical. After every verse, Miss Carter would blow her games whistle with all her might. All this would help the Rescue team find them that much quicker.

They began with 'The Lord's My Shepherd' because it had four parts, so you had to think about it, and everyone knew it. Darryn Perkins had the giggles when they got to the bit about 'the quiet waters by' and then they all laughed. Then they sang 'There was an old man called Michael Finnigan', and at the end of that Miss Carter crept to the back of the tent to find that the Reverend Parkston was slipping away peacefully from life without making so much as a sound. She was much shaken, and struggled for a while until she had got a grip, and could go back to the children with perfect composure. Then, with great gaiety, she suggested they sang 'The big rock candy mountain', with arm-slapping exercises to keep them all warm.

All around them the dark came down, and several of the younger ones, despite the arm-slapping, began to grow drowsy with the cold.

*　　*　　*

Cheryl had finally been persuaded to put on her cagoule, but it was too late. Steven could see she was already very chilled. And she would not budge an inch. They had been here for almost an hour, on the edge of the ravine, where the path had given way and they had just stopped themselves from tumbling into the unimaginable depths below.

It was getting dark. Cheryl had stopped nagging him and telling him he was useless for getting them lost; now she said all she wanted was to go to sleep. Steven had read somewhere that if you were cold and you went to sleep you never woke up again, so he just went on nudging her. As long as she yelled at him every time he did it, he reckoned she was not too far gone.

There was no doubt now that they were hopelessly lost. Steven could not stop worrying about Miss Carter and the poor old Vic, and all the littl'uns stranded up by Ramble's Tarn, depending on him. It made him all choked up with anger and disappointment. He had let them down.

Still, it was up to him now to save Cheryl from dying of exposure or going off the top of a crag, although the way she was now, grizzling to herself, he would be grateful if she would just move! He unrolled the polybag, took a deep breath and shouted at her (she was getting dozier every minute) that she had to climb inside. She refused point-blank and told him to leave her alone. She was hot, she said, and she was going to take off her cagoule, and then she might take off her jumper. Steven did not know whether she was just winding him up (the way she did when she was behind him in the choir stalls, kicking his legs) or whether she had really gone off her rocker.

He climbed into the polybag. Cheryl, sniffing to herself under the rock, was nonetheless sleepily intrigued.

'What you lying down in a flaming dustbin liner for?' she asked, in spite of herself. The words came out a bit garbled, Steven noticed, as though she had been at her Mum's whisky, but at least she was interested.

'Don't you know anything, dumbo? This is a flipping polybag. It saves you from dying of cold. All the mountain climbers have them . . . when they're up Everest,' he added hastily. This last was a stroke of genius.

'Why haven't I got one, then?' Cheryl, suspecting a bit of short-changing, was suddenly more alert.

'You have to share them, Cheryl. Use your body heat. You have to squash in next to me. It's dead warm. Come and see.'

Cheryl, as Miss Carter frequently observed, was a trifle precocious in her ways. She derived maximum enjoyment from making Steven as embarrassed as possible, not to mention digging her spiky knees into his chest. He hated her by the time she had finished, and heartily wished he had left her somewhere in the mist, instead of being marooned here, in thick fog, with a dopy female who was a complete liability. But he knew that Miss Carter would see it rather differently.

*   *   *

It was very dark by now, and the weather had deteriorated from misty clag and cold, to sheets of icy rain and gale-force winds, which whipped at the tent walls. Despite the poor vicar's last need for peace and air, Miss Carter had been forced to jam most of the children at a time into the tiny tent, while she huddled outside with the other half of the group, under an improvised bivouac of anoraks. It seemed terrible to put children in with a dying man, but there it was. Mark Blamire, who was in charge of the matches, had got them wet, so there was no chance of making a fire, even if they could have found fuel anywhere on the desolate, tree-hungry fell top.

And she was getting very anxious about little Emma Thompson, who was not very strong at the best of times. In the last hour she had become very sleepy and white-faced; Miss Carter had had trouble keeping her awake. They had run out of hot drinks and Kendal mint cake. It was a matter of keeping warm by huddling together.

They began to sing again, while Miss Carter wrapped Emma in her own cagoule and squatted by her, rubbing her cold little face. It was a rather tatty effort for a choir of such a normally high standard and the words, 'The sun has got his hat on', were hardly appropriate, although very good for morale. They were doing their best with the third time round when suddenly, out of the grey, driving rain and the dark, a huge wet shape was amongst them; a shape lit somehow by a dim light, with a warm, sandpapery tongue and a deafening bark. It seemed to have a lot of hair and be

161

very pleased to see them. It jumped about among them and then, as suddenly, disappeared into the rain-soaked night.

'Quick!' Miss Carter struggled to her feet, having been temporarily knocked sideways by the shape. 'If there's a dog, there will be an owner not far behind. It felt like a sheepdog to me!'

With one accord, they began to shout at the top of their voices, all except Emma who had finally succumbed to sleep. Miss Carter blew her whistle with all her might. Nevertheless it seemed hopeless. The deafening howl of the wind; the rain crashing against the tentside . . . how would anyone hear them? If only they had a light!

The barking began again, distantly, but growing closer. It sounded like more than one dog. They stopped shouting and listened. Then, much to the children's delight, not one but two dogs, both glowing eerily with that peculiar green light, came bouncing at them through the dark, dispensing joy and affection indiscriminately. There were shrieks of mirth; tiredness and fear were temporarily forgotten. Then, behind the dogs, Miss Carter was suddenly aware of lights, figures, flickering in and out of the rain. She staggered forward, suddenly weak with relief.

* * *

In an instant, it was an adventure! There would be a helicopter, to take the vicar to hospital, along with Emma and one or two others who were poorly (suddenly the object of much envy). The rest would be led down the mountainside. The two Search Dogs, Ruff and Sam, had gone straight off to find Steven and Cheryl. They could not have got far. The team leader reassured the children. There was nothing to worry about. They were safe. What a story to tell at school on Monday.

Miss Carter was more than a little poorly herself. When the team doctor had shone a torch on the vicar's face, she

162

knew. They were not telling her because it was bad for morale . . . but she knew just the same. And Steven and Cheryl were out there somewhere, in these appalling conditions. She had given the team leader as clear an idea of their projected route as her muddled wits had been able to muster . . . but would they be found in time? It was she who ought not to survive this night . . . it was she, Margaret Carter, who was to blame for everything.

But shame and self-pity were extremely negative emotions. Miss Carter swallowed down the tears and got on with the job of preparing the rest of the choir for a trek down the mountainside in the dark, giving them hot drinks from the team flasks, and food; wrapping them in warm clothes. If only they would stand still long enough!

*   *   *

Steven did not know exactly when it was he realised Cheryl was unconscious. He had been talking to her for a long time, and sometimes he thought she answered him, but now he was pretty sure she had not said anything for a very long time. And he was so cold now, that it was hard to think what to do next. He felt very, very small and alone. It would be lovely if his mother would find him now and give him Marmite soldiers and tuck him up in bed . . .

He woke with a start, still dreaming. There was a dog licking his ear . . . and a voice:

'Show me, Sam! Show me!'

The dog was making a peculiar chomping noise, like an old man who has lost his teeth. Steven wanted to laugh. It was a lovely dream. In fact he would close his eyes and dream it again . . .

He was being shaken, hard. There were torches shining in his eyes. The rain, which he had ceased to notice, was pricking at his face. He felt next to him for Cheryl, but the polybag was empty. He was thirteen, and boys of thirteen

163

do not cry . . . it was just that he was sure Cheryl was dead, and it was his fault for not taking care of her. Then he was dreaming again, because there was a terrific noise clattering above him and red smoke and orange lights and the rotor blades of a helicopter which he was seeing from underneath while he was swinging round and round and round . . .

* * *

Steven and Cheryl were both in hospital for several days, suffering from the effects of severe hypothermia. But they were both young and strong, and within a week they were back in the village, to find themselves instant heroes. As for the horror of having had to share a polybag with each other . . . that was something both of them kept 'mum' about. Village children have long memories, and the teasing would have been unmerciful. They were just grateful to be alive . . . but even that would soon be forgotten.

For Miss Carter, there would never be the merciful forgetfulness of youth. Although all the children were unscathed by the experience, and even nagged her to go again next year, she would never forget that she was responsible for it all . . . for however much kind friends tried to reassure her, she knew in her heart that if she had been firmer, if she had not tried so hard to avoid hurting the vicar's feelings, if she had taken different decisions, all might have been well . . .

* * *

There was only one strange thing about the whole affair, from which in an odd way she took a little comfort. She never spoke about it in the village, but the facts could be vouched for by several members of the rescue team. And the following year, it was to appear as a laconic entry, under the 'Incident Reports' for March of that year. As they had made their perilous way down the mountainside, through the clag and rain, she had asked the team leader, who was beside her, how on earth they had managed to find them so quickly, when they were in quite the wrong place.

'After all,' she said, half-shouting into the dark, 'we didn't have any lights . . . not even the primus . . . and we were on quite the wrong side of the fell from the route map I had given Mrs Thomas . . .'

He leaned towards her in the dark, shouting into her ear.

'Sorry, Miss Carter, I didn't quite catch that . . . mind your feet here, it's a la'al bit slippery, now. I'll take your arm . . .'

'I said, we didn't have any lights. How did you find us?'

'Yes, it was your lights, Miss Carter.' His words were half blown away by the wind. 'We had spent four hours on the far side of the fell . . . quite the wrong part . . . getting nowhere, and then a few of the lads saw your lights, like . . . on far side of Ramble's Tarn. Well, your Mrs Thomas, she said you'd be sure and stick to your route, knowing you . . . Anyway, we thought we had better send ower dogs, just in case . . . and it were a good thing we did. We'd never have looked there till morning, and by then you'd have been in a reet pickle . . .'

'But we didn't have any lights . . .' She screamed it into the wind, her voice failing. 'Mark Blamire got the matches wet, and we ran out of paraffin in the stove, and Steven had the only torch . . .'

Miss Carter felt level ground at last underfoot. Her legs gave way with relief. She felt suddenly very weak.

'Steady on, now!' It was that kind voice again, by her ear. The wind had died. There were car lights ahead; voices. They were safe. 'Did tha say tha had nae lights? But we saw them oop on fell . . .'

'It was probably Steven . . . with the torch . . .' she was sleepy now; could hardly keep her eyes open.

'Nay, young'uns were ower back o' fell. Where search dogs found them . . .'

'Well, Mr Nicholson . . . I can't understand it . . .'

'Hey, steady on now, lass . . .' Bill Nicholson caught her deftly as she began to sway. 'Coom on, lads, give us a hand here . . . Ay, it's a rum do about them lights. Our lads definitely saw them. I reckon they saved your bacon . . . a bit of a miracle, you might say.'

Those were the last words she had heard, before strain and exhaustion overcame her, and she knew nothing more . . .

The dear vicar was brought home to rest under the shadow of the fell, in the ancient churchyard of St Oswald's. He at least was now at peace . . . but she would never be able to forget . . .

15

It was unusual to get a search call-out during the day. Mostly, unless it was an accident up on the fell, people waited until dusk was falling before they began to realise something was wrong. But this time, it was while John was at work on a Wednesday morning that the call came through from SARDA. He was given very brief details. As many dogs as possible were to be deployed on the fell above the lake to search for a man who could have been missing for three days.

Social Services were pretty understanding, and he was allowed a small amount of time off for rescue work. He made his excuses hurriedly and sped home as quickly as the lunchtime traffic would allow, to collect Sam and his gear.

Sam was tied up outside the house, looking penitent. He was covered in mud and had obviously just been off on one of his 'expeditions' to the bitch in the village and was in deep disgrace. He could hardly believe his luck when John's car swept into the courtyard, and he realised he was going to be reprieved. Tina appeared at the door, drying her hands on a towel.

'I've just been rubbing him down. Look at him. I can't get it off. I had to apologise to that poor woman again. He pushed past me when the postman came to the door with a parcel. I shall be glad to see the back of him today. He just won't settle. He's had the scent of that bitch in the village and that's it. He can't think of anything else. And to cap it all, he must have gone swimming in the pond on his way there. He really is the limit!'

John undid the rope on Sam's collar.

'Sam, you must learn some discipline. Your mistress has quite enough to worry about without pulling you out of scrapes all the time. Now concentrate. It's time to go to work.'

Sam looked eagerly up at John. Is this what you get for being naughty? he seemed to be asking.

'But you are still in disgrace,' said John. Sam's tail dropped and he put on his most hangdog expression, until John emerged from the house a few moments later, laden with sandwiches and flask and haversack, and Sam's own Search Dog jacket. Then he could not keep it up any longer: he bounded over to the car, tail wagging, and jumped into the driver's seat, just in case he should be forgotten and left behind.

'Thanks very much,' laughed John. 'I've got to sit on that.' He kissed Tina goodbye. 'I don't know when I'll be back. It could be a long job. But Lakewater team are co-ordinating the Search, so you can always ring in.'

They were to rendezvous at the man's house, a cottage high up on the fellside, overlooking the lake. It was a cold, raw day, when the idea of spring seemed a long way off; the sort of day when it seemed a huge claggy could have settled over the whole of Cumbria, when you could not even see the fells on the skyline. They seemed to have disappeared completely.

Within an hour he was there, to find that several cars belonging to SARDA members had arrived ahead of him, and a couple of Land-Rovers belonging to MR Teams. The whole operation would be co-ordinated by Lakewater MRT, and their team leader was there already, poring over a map. There was also a police van, and as John got out of the car and went round to the boot to let Sam out, he saw a couple of policemen come down the track from the end house of a row of cottages. One of them was speaking into a radio as he came.

The team leader of Lakewater, Frank Perkins, came over to John, who was sitting sideways in the driving seat lacing up his boots.

'What's the crack?' John asked.

'Elderly man, lived in the end cottage, last positive sighting somewhere around lunchtime on Monday, when he was seen walking along the fell path, quite close to home. Nothing since then, but we were only alerted this morning when his sister-in-law rang us. Apparently he hadn't kept his weekly appointment with her to do his shopping in Keswick on Wednesday. She was worried he had had a fall, or something. Anyway, they sent up a policeman from the village. He reported the house was empty, everything just so, breakfast things laid, as though he only popped out, like, for a few minutes. Trouble is, he's lived alone apparently since his wife died, so no one knows too much about him.'

One of the policemen came over.

'One of our blokes has just been down to the dairy. The milk is delivered every other day, and he has two pints, so that's about right. There were just the two pints outside, and no post on the mat, so he could have been around till yesterday, or even last night. There's no one living in any of the other cottages now, so just because he hadn't been seen, doesn't mean he wasn't here.'

Frank nodded thoughtfully.

'Well, our team are co-ordinating the Search. I'm calling out all the available dogs. I thought we'd start from here, just by the cottage, and work outwards. Though I'm rather hoping a few more dogs will turn up soon. Otherwise we end up deploying them like peas on a drum and we can't be sure of covering the ground thoroughly. Then we'll bring in all the available teams if we draw a blank. But I want to give the dogs a chance while the search area is clean.' He turned to John. 'Will you come over to the Land-Rover? I'll give Sam the area up from the cottage where there's a lot of scree, above the path. The other three dogs had better quarter the fell to the right and left of the path and, when the others arrive, I'll send them in further down towards the lake.'

More SARDA members were beginning to arrive . . . two of them from the far side of the Pennines must have really pulled out all the stops to get there so quickly, and more vehicles, including Lakewater's transit van, were parked at the bottom of the track.

'I'm not calling out any more team members until we can see what the dogs can come up with . . .' He stopped. The police sergeant from Lakewater village broke in.

'We've just had some further information which could be very useful. Apparently the missing man had a history of heart trouble. He could have had a heart attack and wandered off the path.'

Frank nodded grimly.

'If only we knew for certain when he left the house. Still, it helps.'

John called Sam to him.

'We're off, Frank. Don't want to waste any more time.' Sam bounded ahead of him. They made their way up the track towards the house. The wind was blowing obliquely across the fell. They would have to climb high and work down, he thought. But Sam was already figuring it out, turning every few yards and snuffling the wind. He had picked up the other dogs farther down the track, but they would not worry him. He knew the difference. As they passed the end of the row of cottages, close against the fell, John looked over the wall at the back, a routine doublecheck. A few daffodils had braved the April cold and were a bright splash of colour against the grey stone. This must have been his garden, he thought to himself. Perhaps he'll come back to see the daffodils out like this . . . but he had a feeling, that gut instinct, which told him it would not be so . . .

As he stood there, looking down, he saw a taxi come grinding up from the lakeside road, and stop at the bottom of the track.

'Look, Sam,' he said, 'one of the SARDA members must have come into money . . .' but Sam was too busy doing his job to pay much attention . . .

* * *

The taxi stopped at the far end, which was rapidly being turned into a sea of mud by the number of vehicles coming up it, and an elderly woman, rather flustered, got out and fumbled in her bag for some money for the driver.

The other policeman, Sergeant Baker, made his way down the track towards her.

'Looks like a relative, I would guess,' said Frank. 'Perhaps she can fill us in a bit.'

The woman was still fumbling in her bag. The sergeant was helping her. They spoke for a moment and Frank watched as the taxi reversed away down the track and roared off down the lake road. The woman struggled up the slippery path with her hand on Sergeant Baker's arm.

She arrived at the group much out of breath.

'This is Mr Cross's sister-in-law, Mrs Booth. She felt she ought to come up and see if she could do anything. Come all the way up from Keswick by taxi.'

'There isn't another bus today, you see.' She pulled out a hanky and began mopping her neck. 'I hope I did the right thing calling the police. I thought, I'll feel such a fool if I've got it wrong. What if he's just forgotten the day, and he's back home by the time the police arrive. But we've been worried about him for a while. His heart and everything, and he's been so depressed since my sister died. We tried to get him to come and live with us, but he wouldn't have it. He said they'd always been here, since they were married and he'd be deserting her.'

'You did the right thing, Mrs Booth.' Sergeant Baker patted her arm. 'If you can spare us some time for a crack, you can be very helpful to us. We've got to fill in the picture of his last movements. One of my men has gone down to the cottages along the road, to ask questions, but we get the impression he was rather solitary. Kept himself to himself, like. Would that be right now, Mrs Booth?'

The old lady nodded energetically.

'Neither of them wanted much company. They were funny like that. Towards the end, when she got bad, they had the television, but he didn't like it much, you could see that. But he nursed her wonderfully: I couldn't ask for better. She were quite bedridden at the end—the last three years—and it's funny, thinking about it. I' was only wondering the other day how he was getting on. It was the anniversary of her death just a few days ago. Now he didn't seem to take it too bad at the time, but you never could tell with him. And a few weeks ago when we saw him in Keswick, he looked terrible. He didn't seem to know us. We took him for a cup of tea in the Wimpy bar, brought him round, like. It was as though he were dazed. It gave me quite a turn . . .'

*　　*　　*

For many years the bus had stopped at the bottom of the track, and together he, and sometimes one or two of the others from the row of cottages, had got down, waved goodbye, and climbed away up the fell. It was not an official stop, of course, everyone understood that, but as long as there had been people there, the bus had stopped, and that was that.

But then the couple in the end cottage left and went to live with their married daughter in Preston, and the old man in the middle cottage got taken away to a home in Carlisle, and very soon they were the only ones. Then the bus company did some 'time and motion' study (Jed knew all about that, being a carpenter for the council), and all 'unauthorised stops' were forbidden. So the bus no longer stopped at the end of the track, its windows lit up and fusty—a good sight on a winter's night even if you were not on it, made you feel comfortable, his wife said.

It was things like that which cheered her up. She could not get about the way she had, and while he was still working he was glad of anything which kept her mind off the pain in her legs—even the television, which was a mixed blessing in other ways. But of course doing without the bus stop was a blow, adding another twenty minutes to the time before he could get in and fix her tea and make sure she was all right. It was a worry now the neighbours had all gone.

Anyway, they managed. He got up before dawn and, while the sun rose over the distant lake, he put out the breakfast things and laid her a fire with a bit of newspaper poking out so that, even with a walking frame, she could light it all right without falling over. Then, by the time he had got her dressed and taken her tea, it was time for the walk down the lake road to the official bus stop, and another long day.

But they had been so happy. He would not have given a day back. The fell was always there, wild and free, winter and summer, behind the house. You could feel it in the night, comforting. At least that was what she said. And the lake, catching the morning light like that . . . it took your breath away. You could not have a better place to live, and they still had each other, and in a year or two he would retire and they would be together all day, and he would never have to worry again about leaving her to cope by herself.

But it had not worked out like that. She had left him alone . . . just when he was planning how they would always be together when he retired, and how they would celebrate their Golden Wedding here in the cottage where he had brought her as a young bride. She had left him, one long, terrible spring day when the late snow lapped against the lake edge and blocked the track, so that, for the first time in fifteen years, he could not get to work, even if he had wanted to. And that day he had stumbled through miles of drifts to the nearest house on the main road, and the doctor had come up, and when they had cleared the track they had taken her away in an ambulance with pneumonia, and she had never come back alive.

He could not forgive that. She had never seen the daffodils he had planted for her in the small garden on the side of the fell, just under her window so she could look out; she had never seen the new shelves he had put up for her, just at waist height, to while away the time she was in hospital; she would never see the rainbow over the lake again, the way it seemed to go through the little white house on the far side. It just did not seem right.

He went on working, though. He was glad of it, and often put in hours of overtime so that he would not have to go back to the cottage, with all its memories and no one left to care for. And still he got up before dawn, winter and summer, and laid the fire just the way he had when she was alive, and the breakfast things, and left everything spick and span. But it was not the same.

Then he began to get the pains. Once or twice he came over queer while he was at work; it was heavy work sometimes, lifting big planks of wood, and a lot of bending. The council sent him for a check-up.

They called him in one day. They were very kind. They were retiring him early, on health grounds, they said. He had a bad heart. He did not believe it. Mr Kitson, his boss, said he could take it easy now; do his garden. But Mr Kitson did not understand.

There was nothing else. It was not as though he had ever been idle. But it was autumn when they laid him off and he had the whole long winter to live through, with the dark coming down by four and the long, long evenings with only the television yacketing away; he could not stand it, even though she had liked it.

He worked in his garden and did things in the house, but now there was no one to show it all to: it seemed pointless. Every Sunday he would take the long walk, in his best Sunday suit and stout shoes—up the fell path to the far end, where his nearest neighbour lived, and out by the gate and down the lake road, in a big circle. He tried to tell ·himself it was a good thing. He had not walked for years, had not wanted to leave her alone. Now he was free. He could go into Keswick shopping. His sister-in-law was always nagging him to. He could listen to classical concerts on the radio. He could play his violin. She had never liked what she called his 'caterwauling'. But what was the good of being free, if there was no one to share it with . . . ?

That Christmas he went to his sister-in-law's, the way they always had. But, of course, it was worse. You could be alone in a crowd, and even among family. He had not seen it before. But there was a great hole in him where he had used to feel. Now he dared not feel anything any more. Not that she was unkind . . . Vera wanted him to come and live with them. But he could not leave the lake, and the home they had made . . . and the daffodils he had planted for her . . .

She had been buried in the lakeside church on the edge of the village on a black Saturday in April, when the daffodils were coming through. But the earth was black with rain and there were black clouds on the tops, and even the fell was angry. But at least she had died here, close to home, and perhaps she had seen the fells before she died, out of the window of the hospital, and knew she was coming home . . .

* * *

John had climbed very high by now, up onto the western approaches of the fell. The clag had been blown away, and now there was that brilliant April light which lit the fells like a stage set. Hard to believe they were real. Hard to believe that somewhere, perhaps behind the next boulder, there could be a body, lying there.

He was sending Sam away to his right so that he was working diagonally, covering with his nose the whole littered area of scree, which some long-dead glacier had left in its wake. He wondered to himself how many men they would have needed to cover all the caves and crannies and boulders, before they could be sure. But a dog, working properly, could eliminate huge areas with a sweep of his nose. Still, he had to be certain.

He stopped to get his breath. The wind caught him. It was free and laden with icy hail. There was something in it which made his blood race . . . some clear, cold scent, as though from another world. He looked across the lake. There, just visible beyond the nearest fells, were the highest peaks, tops covered in snow, the sunlight catching them to gold. Wonderful to live up here, even with all the loneliness.

He was climbing by a beck which ran down almost against the cottages. Now, after the spring melt on the lower fells, it was in spate, its whitened spume hurling itself angrily at the black rocks. It was far too high up here . . . a man with a heart condition, getting on a bit. Would he have come this high? It was illogical, but so much depended on his state of mind. John decided to go on up a little way.

Sam was in front of him suddenly. Feet planted firmly. He wasn't 'indicating', just looking at John in a puzzled way.

'Show me, Sam. Show me.'

It was unlikely to be a body. After all this time, they could read each other so well. It was just something Sam thought he ought to check. Sam was running ahead of him, up the fellside, close to the little valley, where he had already searched. John 'read' him again. It was not that excited, self-important waddle which heralded a find. It was something else.

Sam, paws planted with precision, was making his way down the steep side of the beck, tail wagging. Then, with one bound, he was in the middle of the rushing water, all four paws scrabbling at a slab of slippery rock at the head of a small waterfall. John scrambled down the side of the bank, half-sliding on his bottom. What was Sam up to now?

There was a small, mossy overhang on the far side of the beck, where a little tributary of the main waters would normally have trickled down, a fine sound on a hot summer's day. Now it was its own miniature raging torrent; but Sam, oblivious to all such distractions, made another leap, onto the far side of the main beck, paws scrabbling furiously for a foothold, and disappeared under the overhanging ferns and plants into a miniature cave.

'And how do you expect me to get over there?' John shouted above the roar of the torrent. It was a rhetorical question which Sam ignored. Only his tail was visible, waving madly as he pursued ghosts of his own.

'It had better be important!' John cursed as he got a bootful of icy water trying to find a safe place to cross the stream. But Sam had not finished. His head emerged regretfully and, without a backward glance, he scrambled on to the top of the overhang and began to nose around in the bracken cover above it.

By this time, John had managed to ford the beck and followed him up the rock. Sam was still nosing about above the overhang. John wondered how, logically, he had managed to pick up whatever scents he was finding when he was theoretically working 'backwards' against the wind . . . but sometimes it seemed as though he had developed other senses . . .

John stuck his head into the miniature cave. An overwhelming stench hit him and he recoiled, staggering out into the fresh air. Inside he had just glimpsed a huge bloated shape lying against the far wall.

'Sam. What the hell are you playing at? A dead sheep! I thought you had grown out of that years ago. Getting my feet wet for a dead sheep. Wait till I catch up with you.'

Sam was barking at him. He was a few yards further up the fell, tail going. John scrambled after him, slipping on the wet grass.

There, tangled in the still-barren branches of a rowan tree which overhung the beck, was a plastic bag. Very new and shiny. Bright red. The colours not bleached by the weather. It could only have been here a few days. He would have spotted it if they had been walking on the western side of the beck. But as it was, Sam knew something. He might have been momentarily distracted by the dead sheep, but he knew there was something else. Once again John felt he had been taught a lesson.

He climbed up, and carefully extracted the bag from where it rested. There was a name on it ... 'Thomas Atkinson, Nurseryman, Carlisle.' He pulled it open. Inside was nothing but what looked like a handful of peat ...

Sam stood watching him intently, his tail still wagging at the tip. Had he done all right?

'Well done, Sam. Good boy.' Sam pranced around his master, looking pleased. He was glad now he had ignored the dead sheep and come on up the fell, although it had been hard, leaving something so deliciously intriguing. Still, he knew all about dead sheep. They were all right to investigate if you were not working ... but he was on duty now.

'I don't know if this has anything to do with the search, Sam,' John said, 'but we'll mark the place on the map and tell them over the radio.'

He pressed the button on his set.

'Lakewater Control from Dog John . . . Have found a plastic bag. Looks as though it hasn't been here very long. Sam was very interested. Could have been a walker . . . an address in Carlisle, some nursery. Seems to have had bulbs or seeds in it. Peat, anyway. Hard to tell.'

'Search Dog John from Lakewater Control. Please mark exact location of find and work upwards from there, in case there is any connection . . .'

Sam had forded the beck once more, and was sitting, waiting. John had not noticed that close by, on the beck's far bank, close to where Sam waited, the ground had been disturbed. He struggled once more through the white water, earning another bootful as he went. Sam watched him, tail thumping. John sometimes wondered if he was amused by his blundering human with his inadequate coat . . . after all, not everyone could have a coat like Sam's, dense, thick underlayers which never really got wet, and which insulated so efficiently. Humans, with their clumsy boots and their heavy-weather clothing, were so badly equipped for survival! You found that out when you fished them off the fell, half-dead or worse. Pitting themselves against the savage elements, time after time—perhaps that was what attracted them to its wildest places, after all . . .

Sam had been right. The moss on the far side of the beck had been dug up. He knelt down and probed with his fingers. A few centimetres down he encountered a smooth shape. He eased the earth away. It was a bulb! Someone had been up here—had come up here—to plant something on the fell . . .

They worked on upwards until they reached the summit, far up above the little row of cottages nestling under the shadow of the fell. The short northern afternoon had long ago faded into night. John had seen the distant snowy peaks touched to fire by the sunset. There was snow here, too, in the high places, lying like shadow in the corries. It

179

was the last snow of winter, but if they had more—and it was cold enough for that—a body might not be found until full summer.

But in their own little patch they could be sure there was nothing—every cave, every granite boulder, every patch of lethal scree slipping under the wet, had been covered by Sam's questing nose. In the distance, far down in the valley, John caught glimpses of the line search, with team members from five MRTs fanning out over the intake fields. It had seemed a good idea to look in the area around the little churchyard where the lost man's wife was buried. Bursts of cross-talk from the radio told him that the other Search Dog teams had also drawn a blank. Frank had called a full panel search which involved all the men and dogs from teams attached to the Lake District panel. There was another sweep search taking place on the far side of the fells, among the steep crags, out of reach of the radio. Perhaps they would find something there . . .

When, eventually, they reached the foot of the fell, it was after midnight. They were both soaked to the skin, the clag having come down again with a vengeance. The whole area was a mass of Land-Rovers and vans and private cars. Team members were crowding round Frank and the Control Land-Rover with its tall radio mast.

A wonderfully welcoming smell came drifting up at them through the mist . . . onions . . . that meant hot dogs . . . and soup! Sam lifted his nose and sniffed. Not perhaps quite as good as dead sheep, but no doubt his master would appreciate it more! The WRVS had arrived with their mobile catering unit—and were rapidly being eaten out of supplies by ravenous searchers, some of whom, like Sam, had been on the hill for ten hours or more.

John was depressed—even after a couple of hot dogs and a cup of scalding soup. He could be sure they had covered their patch thoroughly, that there was no body—alive or dead—on his part of the fell, but to find nothing . . . all of them were feeling low. They all knew in their hearts that after all this time there was no chance of finding anyone alive. They took refuge in jokes, or in half-hopes. It was always hard to take.

'He's probably taken off to Blackpool for a week, without telling a soul . . .' said someone gloomily. 'Having a whale of a time while we're slogging our guts out oop here.' It happened sometimes, that kind of thing. All too often. But some instinct seemed to tell them different this time. It had all the ingredients of tragedy.

Sam slept on in the back of the car, on his special rug. Before John had had anything to eat, he had rubbed Sam down with a towel and given him some food. It was important to look after the other half of the team.

They fell to discussing the plastic bag, and the bulbs planted by the beck. It was the only positive clue gleaned after a long night. The line search on the far side of the fell had turned up a dead bullock beyond the stone wall, and the farmer had been informed. But nothing else.

Frank shook his head in disbelief.

'The only thing puzzles me: a bloke who would plant daffs out on fell to make it beautiful would hardly leave a socking great plastic bag for some poor sheep to choke ower . . . not unless he was that confused he didn't know what he were doing . . .'

'Do you think it's the bloke we're looking for, Frank?' one of the handlers asked. 'The bloke who left the plastic bag, I mean?'

'We have to keep an open mind, that's what I say.' Frank shook his head. 'I've been trying to get inside his head, like. After what the sergeant told us. Seems he's had a bit of bad time lately. Enough to upset anyone. But who knows at this stage . . . ?' The radio crackled.

'Lakewater Control from Peldale One.'

'That's the line search down by the church,' Frank said. 'I wonder if they've come up with anything.'

'Peldale One from Lakewater Control. Go ahead. Have you found anything, Bill?'

The familiar rich tones of Bill Nicholson came floating out of the radio. Frank moved it a couple of inches further away from his ear.

'. . . Nowt down here. We've been all ower ground and we're aboot buggered. Can we coom home?' Frank laughed.

'. . . Yes, Bill. Come in for a briefing and get some hot food. I suppose you can smell the onions from down there . . .'

He came round to the back of the Land-Rover.

'Right then, lads . . . and lasses,' he added as Ann Carter, with her dog Barney, appeared out of the darkness trudging wearily towards them. 'I think we have done about all we can do tonight. Covered all the immediate possibilities. We'll have a full call-out again for all those who can make it—for first light tomorrow. As many SARDA members as possible. We shall have to extend the search into the surrounding fells. Now go home and get some sleep.'

'They'll have the police divers out in the morning,' said Ann gloomily, as she towelled her Collie down. 'He must be in the lake. It fits, and we've searched everywhere else . . .'

* * *

It was a better way to remember her than any headstone in a churchyard. He did not believe she was there any more, not the real person he had loved. She was up here, in the high fell. Her voice seemed to speak to him from the beck, calling to him. He would walk miles, thinking about the times they had had together. It eased the pain. It was the only time he found peace. And the thought that in the years to come, when he, too, would be gone . . . the whole fellside would flower with his own memorial to her. There could not be a better way to remember . . .

'He were always a quiet sort of chap. Didn't talk much, but allus polite, you know what I mean. And I hear he was right good to that poor wife of his . . . she as was crippled. Couldn't have been kinder . . . After other cottages were empty, like, we were his nearest neighbours oop on fellside. One time he used to coom to us for his milk, before they started leaving them bottles down by the road. Course, I'm going back a bit. Still, he knew me and he would allus pass the time of day. And when I used to go oop onto fell checking sheep, I would pass him on the path. He allus walked the same route: straight down path to my farm gate, and then round by road, down by the lake. I believe he used to give himself an hour away from his lass, and only on a Sunday, so she wouldn't worry, like. But after she died, well, that were different. You nivver knew where you might meet him. Ay, that were a strange thing. I caught him once, right ower back, miles from home. My dog started barking, and there he was. He looked to me as though he were digging summat oop, or planting summat. But I was that embarrassed to ask him. He got a bit vague like. Well, poor bugger, he didn't have much left . . . not to hold on to, like . . .'

*　　　*　　　*

The full panel search went on for three days. By the end they had covered, with the help of five teams and eight SARDA members, the whole area of fells within a five-mile radius, up to the summit of Bell Pike at over two thousand feet. But there was nothing.

By now the possibility of a body being found in the lake was very strong. It was a deep lake, curving round behind Bell Pike towards the great bulk of Tanner Howe in the east, with Muddlewater lake strung out beyond it. The divers and searchers concentrated on the nearer shores of Lakewater, but they, too, found nothing. A body would normally take between ten days and a fortnight to surface, brought up from the bottom by its own gases. So perhaps they would have to wait.

Frank was planning to scale down the search at the end of the third day. Searchers and dogs were exhausted and depressed, as always at the end of an intensive but unsuccessful search. It was hard to keep morale high. The dogs were feeling it as much as the men and women on the team. It was time to wind down. They had done all they could.

Then, as if to place a full stop firmly on the operation, snow began to fall late in the afternoon of the third day. John and Sam were covering the area below Bite Tarn for the second time. They had searched a major part of each day, with breaks while John dashed in to work to delegate as much as possible to an understanding team.

He had just decided that, as he and Sam were at the farthest point of the search so far, he would go down into the far valley and allow Sam to quarter the area of broken scree and granite outcrop on the western face of Scarpdale, which looked over Muddlewater lake. It seemed unlikely that the man could have possibly strayed what was such a long walk away from the cottage. John had left his car at the bottom of the footpath which led upwards towards Bite Tarn—a popular tourist walk in summer.

Within minutes of feeling the first flakes stinging his cheek in the wind, Sam had come sneezing and huffling against a full blizzard, the light had gone and they could not see a thing.

The radio crackled. The blizzard had hit the main body of the searchers a few minutes before. By now it was a complete white-out. Frank was calling in all the teams off the fell.

John and Sam struggled back down the path. It was

184

astonishing how quickly the fells were transformed into white, unearthly places. And the silence, always deep, was, in the intervals between the screaming blizzards which whipped them mercilessly, suddenly profound, endless, frightening in its emptiness. John was glad to see the car ahead of them in the fast-gathering darkness. It was a comforting sight, even if the bonnet and roof were now covered in at least a couple of centimetres of snow.

John thought of the body which might still be lying out there, somewhere on the bleak fastness of the fells. Very soon it would have its own burial, which would keep it safe and inviolate from crow and buzzard and fox. Early April snows often came deep, and might stay till May on the high tops. They would search again with the dogs, perhaps when the conditions had cleared, but there was little chance now. Only if the body was in the lake would it be found.

*　　　*　　　*

But no body ever emerged from the lake, to solve the mystery. It would be a long time before the last few pieces of the puzzle would, unexpectedly, fall into place, and it would only be after a near-tragedy, that John would find the answers to several questions of his own . . .

But before that, there would be a great deal of searching to do . . .

16

The summer came at last, although, to read the Incident Reports for June and July of that year, it was hard to believe that summer had visited the fells at all . . .

In June Sam and the dogs from neighbouring teams assisted Peldale MRT to locate a casualty on Garkle Crag, in gale-force winds, rain and poor visibility. Four days later, on Langton Raise, dogs, including Sam, spent the whole day searching for two men under similarly appalling conditions. They were found that evening by the police, supping a pint in the next valley, unaware of all the fuss they had caused. On that same evening, in shocking weather, Sam and all the other available SARDA dogs were called out to the far side of the Pennines to search for a mentally handicapped adult missing near a local tarn. Much to everyone's relief he was found, safe and well, by two of the Collies—Ruff and Ben—on the edge of Parkin Tarn Moor, just as night was falling.

And so it went on . . . The following night, six dogs were called to Waterdale to search for a missing fell runner. He was found by Pammie, a Labrador bitch who had just joined Bess in the Marrowdale team with her handler Jeanette. The man was badly injured, and Jeanette, who was the local doctor, proved her worth once again.

Two days later, in pouring rain and biting wind, four dogs searched all night for a missing deaf boy, who was found next day, having fallen from a crag. He died later in hospital and, the next day, three of the dogs continued a search of the adjacent fell for six cadets reported lost.

When they were eventually found (in a totally different location from that given by the informant), one had already collapsed from exposure, and the other five were rapidly approaching the same state.

But then, at the beginning of August, when Tina had resigned herself to the garden as a sea of mud, and the children in their wellies for ever, they got up one morning and, looking out of the window, could see, for the first time in weeks, the fells standing proudly against a bright blue sky with not a cloud in sight. It was as though it had always been like this, and there had never been rain or hail or snow in June, as though from now on everything would be perfect.

A soft, warm wind blew across the garden. The wellies were put away in the back of the cupboard. It was the beginning of a heatwave.

If someone had told John that those streams which he had fought through on the fell only two weeks before in angry spate, would now be reduced to a mere trickle, that the level of the reservoir at Thirlmere would drop so far that it would show a thick, chalky tidemark around its edge, that the sheep would stand panting in the middle of the road, and that they would be bringing casualties off the fell suffering from heatstroke, he would not have believed them.

The weeks of baking days and hot, restless nights went on. It was the heat itself which became a problem. Hordes of visitors began to clog up the roads into the Lake District, all desperate for cool and space and fresh air. They climbed away from the overcrowded lakeside towns, in their fashionable shoes. The paths, already worn by countless feet, were carved into deep gullies. In the valleys, carbon monoxide fumes clogged the narrow streets of what had been, in the winter, quiet market towns. On the steep roads up into the fells, coaches and cars fought bloody battles over rights of way.

The Mountain Rescue Teams were out every day. There were broken ankles, and heart attacks, and people simply benighted and lost after a good night in the pub and a wander up into the fells. There were the serious climbing accidents, and the idiots. And every day the teams had to contend with the heat and the crowds. Even a flashing light on the Land-Rover had trouble carving a way through the ice-cream queues down by the lake . . .

It was a time when it seemed that the lakes and fells would sink forever under the sheer weight of numbers. Stand on any fellside and, as one looked across the valley, there would be the glittering metallic snake of cars crawling up the Kirkstone Pass, and the distant sound of revving engines, where before there had only been the cry of the buzzard and the call of sheep on the fellside.

But for the Mountain Rescue Teams it was all part of the job. Sam and John and the rest of the SARDA members were much in demand. Walkers who did not know the fells, and who would not have dreamt of consulting a map to get from A to B, were often unreliable witnesses when it came to explaining where their friends and relatives might be lying injured on a fellside. The search capacity SARDA gave their individual teams was very welcome.

But sometimes John longed perversely for the cold, crisp winter days . . . even for the snow and sleet. Plodding for hour after hour through the intense heat, Sam's coat dark with sweat, was enervating and desperately tiring. Only at two thousand feet would there be a mercifully cool breeze from the valley, but even here, nature seemed dead, the bracken sun-baked, the fells somnolent under the glare of the merciless sun.

It was in the hottest week of the summer, when the Lake District had had no rain for a record-breaking number of days, and the Water Board were banning hosepipes and the washing of cars, that SARDA were called out on a panel search . . . over the Pennines, in an area owned by the

Forestry Commission. The local team had already spent two days searching the wooded slopes of the fell without success. Now six teams had been called in from all over the panel area, and SARDA was requested to send as many dogs as possible.

The car had been spotted in the car park several days before. It was a tourist area, after all, with nature trails through the forest, but usually, long before dark, the car park had emptied, and people had begun to wend their way back to their hotels and guest houses back in the Northumbrian villages below. But the car had been parked unobtrusively since early morning, down the far end, well out of sight of the warden's hut, as though whoever had parked it there would prefer that it would be as long as possible before it was noticed.

It was still there late at night, and the worried warden immediately reported it to the local police. The sergeant came out from the next village and took details. Perhaps it had been stolen. He went back to check with headquarters.

Within a short time he was back. The owner of the car had been traced to an address in Newcastle. The previous day a missing person report had been filed in the city from a worried relative. The man concerned had not reported for his job that week. He had been in a depressed state. He could not be contacted at his home. The police had been informed.

The local team had searched for two days, but had found the condition of the forest very difficult. There was a great deal of windfall, which made a line search almost impossible because of fallen trees and uneven ground. The team had searched the established paths, but there were vast areas which they had had to leave untouched. It was more a job for the dogs and their handlers than anyone else. Only they could go into heavy cover and, with their ability to pick up air scent, stand a chance of finding a body in such a tangle of undergrowth.

Sam and John were briefed by the Controller, and sent into a dense area of evergreen forest, well away from the major paths. It was a rather neglected area, with fallen trees blocking what small paths there were, dead branches and waist-high bushes threatening Sam's eyes, and making walking risky for them both. It had been pleasantly cool when they began but by mid-morning the heat had started to build up unbearably. They were in semi-shade, but the ground, saturated for weeks by rain, and never fully dried out by the sun, created a heavy humidity which was more like being in a tropical jungle than anything they might have expected hundreds of feet above sea level in the northern fells.

But the worst nightmare was the flies. They had been warned before they started searching that they might be a problem, and someone had doled out proprietary fly repellants for everyone. But very soon it was obvious that these were going to be no good whatsoever. The flies clustered densely round John's head and crawled in his beard, hovering in a whining, buzzing cloud just above his head. They tormented Sam by clustering round his eyes and round the soft parts of his mouth so that his concentration was permanently interrupted by having to shake and paw at his head, in an attempt to free himself from them.

All day they searched through the forest with the flies making their lives unspeakably miserable. John had gone back once for an anorak and had been forced to put it on, tying up the hood tightly to try to keep them out of his hair. Sam had been sprayed with fly repellant, but there was little they could do so near to his eyes. Yet still he struggled and scrambled through the undergrowth, gamely pushing his way onwards, his head raised as he tried to pick up scent on the heavy air.

They were glad of the WRVS as the search went on. Where before they had given out welcome hot soup and rolls from their mobile canteen during a freezing blizzard, now the needs of the searchers were very different. An hour or two in the overpowering heat of the forest, streaming with sweat, and both dogs and handlers, as well as all the MRT members pounding their way across the open ground and along the forest rides under the direct sun, were becoming dangerously dehydrated. Cool drinks, water for the dogs, and breaks for sandwiches and cake— although no one felt much like eating—made it all bearable.

The next day was equally hot. By the third they were down to half the number of dogs. It was difficult for handlers to take so much time off in the middle of a week. But the forest covered a huge area and, despite the fact that line searches, and the dogs and handlers, were working through every part of it and out onto the open fell, and dogs were being called in by the line searchers to go into cover which was impossible for men alone to clear, there were still areas, away from the forest rides, which were almost impenetrable.

Sam and John worked on, hardly seeing a soul. Sam's muzzle had been scratched by brambles, and the flies clustered round the wound, making him miserable. John thought longingly of the high fells of the Lakes. To stand on a windy crag, cool lake water below, even on the bitter winter days when snow hampered their every move, seemed to him at that moment the best thing in the world. He swore to himself he would never complain again!

By the end of the third day they were both totally exhausted. All the teams had drawn a blank and were dispirited and concerned. If there was a casualty somewhere in the forest his chances were getting slimmer every day.

But for a full week in the killing heat, dogs and teams carried on searching, although the level of searching was wound down. There were other, urgent calls on the teams. Lost walkers, heart attacks, a missing child out with her parents for a picnic on top of the fells. All the problems of the summer season. But in between other call-outs and trying to catch up on work, John drove out when he could after he had finished for the day, the humidity hitting him like a blanket as he entered the forest, the flies just waiting to swarm around him in a thick, unbearable cloud. At length, it was decided to call a halt. All the areas had been covered by either line searches or dogs. There was no point in going on.

But like all the unsolved searches, the questions nagged away. Where was the body? How had the car come to be left in the car park? Anxious relatives waited for news, but there was none.

John was chatting about the search to Peter Reeves, a friend of theirs in the village, late one evening the following week. He knew his friend had a reputation as a bit of a clairvoyant . . . But that sort of thing seemed to have little place in a modern world of highly trained dogs and sophisticated radio-controlled searches with high-powered equipment.

Nevertheless, his friend suggested he have a go at trying to locate the body. John half-intrigued, went out to the car and brought in a folder of maps of the Lake District and the North. He rooted about and pulled out a large-scale Ordnance Survey map of the forest area.

'Have a look at that, Peter. See if you can give me some hints where else we could search . . . anywhere here.' He indicated with his finger the whole of the forest area and the fell behind it. Peter nodded and then, much to John's surprise, grasped his wedding finger and pulled off his wedding ring. He went over to his wife's sewing basket on the sideboard and brought out a reel of cotton.

'What on earth are you doing, Peter? It looks like a whole load of mumbo-jumbo!'

Peter grinned back.

'Probably is! All I can say is, it's a bit like water divining. Just wait and see.'

He tied the gold ring onto a piece of cotton and,

spreading the map on the table, swung the ring over the map, round and round, an expression of intense concentration on his face. Then he marked a cross in the middle of the forest area.

'There's definitely something there, John.'

John smiled and shook his head.

'I'll believe you, thousands wouldn't,' he said. 'But go on, I'm fascinated.'

His friend swung the ring again. It certainly seemed to go round and round over one particular spot, but John was a trifle sceptical.

'. . . And there's something there, too.' He stopped and made another cross with the biro. 'I don't think it's a body, but I'm not sure. There's definitely something.'

John nodded.

'OK, Peter. Now I'm going to play a dirty trick on you. I've got another map, a Forestry Commission one this time. It looks quite different. I don't think you'll be able to pick out the same places again. But have a go . . .'

They went through the same ritual again. John, having searched the area for a week, knew which part of the map was which, but unless one had intimate knowledge of the forest, it would be impossible to match the two maps. But to his amazement, Peter leaned over the map and placed a cross in exactly the same spot as his very first one. Then he nodded.

'There's something there . . . a body. Probably. You must look there.'

'Oh, come on, Peter. You must know the forest to be able to do that. You're having me on!'

Peter shook his head.

'You know me. I'm not a walker. I've never been near there in my life, unless I was a babe in arms. Anyway, I'll have one more go. See if I can pinpoint one more spot.' He leaned over the map, his face set in a frown.

'Yes. There's something just there. I can feel it. Not a

body. Oh, I don't know. Maybe a body. I can't pick it up clearly. But something, anyhow.' And once again, he took up the biro and drew a cross in the same place as his second mark on the Ordnance Survey map.

John, slightly unnerved, burst out laughing.

'OK, Peter, I give in. It's very clever. But I must tell you, both those areas have already been line-searched. There can't be anything there. We've been all over that area of forest with a fine-tooth comb.'

'With dogs? Have you been there with Search Dogs?' Peter looked serious. 'I'll tell you something: wherever the body is, and there is a body somewhere around there . . .' he indicated the two crosses with his finger, 'it's a suicide, and it's lying in open ground, in the hollow made by the roots of a fallen tree. Like this . . .' he showed John with his hands. 'So a line search, going forward . . .'

'Wouldn't see it . . .' John broke in. 'Yes, I see that. If dogs had been in there, they would have picked up the air scent, but unless the searchers looked back into the hollow, they would never have seen the body. And God knows, it was difficult enough to see anything. I've never known such difficult terrain. It made Helvellyn in a white-out seem like a Christmas party. I'd rather have that any time!'

A couple of days later, when a work visit had taken him to a village on top of the Pennines, he decided to test out Peter's theory. He had brought Sam with him that day, and they drove down towards the forest, the evening cool just beginning, taking the worst out of the day. Sam, sensing that there was work to do, sat intently on the front seat, his tail thumping. The car park was empty, apart from a few stragglers, most walkers having gone home. John let Sam out of the car and, using the map, followed the main ride along to the farthest point of the forest. The flies were no longer so troublesome in the evening, and a faint breeze stirred the tops of the conifers. John wished it had been like this all the time.

He directed Sam away from the path, through the rosebay willowherb, already now wisped with seed heads, into the deeper part of the forest. No wonder so many fairy stories featured this sort of country. This was for the most part new coniferous forest, resented by many as a blot on the landscape, but containing within it the ancient heart of the northern forests of long ago, where wolves had roamed the wild fells and the Romans had fought the Britons almost to extinction.

They struggled through the undergrowth, and came suddenly to an open part of the fell, unexpected, a break in the forest cover where the contours showed on the map a sharp increase in height. Peter had said the body would be in open ground. They began to search. Sam, working in the still air, quartered the ground over and over again in the heat, until, once again, despite the apparent cool, they were both sweating. There was nothing.

The sun had gone down and it was deep dusk, night already under the trees. They were both very tired. John remembered the second cross on the map. If the forest gate at this end were open, he would go through and search the second area before it got too dark.

They made their way back to the ride and John walked along it to the forest gate. But the warden had been out before him, and it was firmly locked, the ride beyond it glimmering under the pale evening sky, quiet and empty. A fox barked somewhere in the dense cover. A rabbit skittered across his path, and Sam made a lunge for it before he remembered he was still on duty. He came back to John, his tail wagging.

'I think we'd better call it a day, Sam. We've done our best. I think we've ended up with egg on our faces. Comes of trusting to alternative technology. Still, it was worth a try.' Sam shoved a hot, dry nose into John's hand, as though to show agreement. 'Back home. Tina will be worrying. And you could do with a drink. We'll come back and search the second spot . . . maybe tomorrow.'

But the next day, no sooner had John got home from work than there was a call-out from the Langenhow team . . . not involving Sam, but a request to help with a stretcher party to bring a heart attack victim down from the fell somewhere above Garth How. And the next evening there was a SARDA call-out to search for two hikers, overdue at their hostel, thought to have been walking from Keswick to the village on the far side of Blencathra. There was no chance of going back to the forest and, after drawing a blank the first time, it hardly seemed worthwhile.

It was after the team had met together at the base of the fell, and were sharing a thermos of tea, that one of the SARDA handlers volunteered the news . . .

'Hey, did you hear? They found the body in the forest. I heard it on Radio Cumbria. I don't suppose they could reach us to tell us this evening, as we were on the fell, but I heard it as I was driving up in the car . . . A suicide, they suspect. Two blokes orienteering just stumbled on him. Must have been quite a shock.

'Where was he, for God's sake?' one of the team members asked. 'I thought we went over every square inch with a fine-tooth comb?'

'In that place? You must be joking. Anyway, they didn't go into details, but it was somewhere in the forest.'

John drove home in thoughtful mood. A few days later, the SARDA dogs were called out to the top of the Pennines to help in the search for two missing girls reported not returned from a walk along the Pennine Way. When the girls had been located and brought down safely, one suffering from heatstroke, he caught sight of the team leader who had been the Controller on the forest search, dismantling the Land-Rover radio mast. He went over.

'Hey, Dave. Can you spare a moment? Whereabouts did

those orienteers find the body we were looking for? Was it actually in the forest?'

Dave opened up the passenger side of the Land-Rover and rooted about for his maps.

'It was rather unusual, John. No wonder we didn't find him . . . They'd already had a line search through there . . . but he was lying in a hollow made by a fallen tree. Hid himself, I would guess, on purpose. I suppose the dogs would have scented him if they'd gone in . . . But walking forwards, as they do in a line search, they wouldn't have seen him.' They craned over the map, following the pointing finger. 'He was just there . . .'

John felt a faint shiver down his spine. He remembered the second cross which Peter had placed on the map. He remembered his words . . .

'Thanks, Dave. I was curious. It was just a theory I had . . .' He said goodbye and walked slowly back to the car, where Sam, hot and tired after an afternoon on the moor, was waiting for him.

The forest gate had been locked that day . . . otherwise would they have gone on, to the place which Peter had marked on the map, and would Sam have found?

Coincidence, perhaps? He would never know . . .

17

It was a warm evening, later in that long, hot summer. John
had just come home from work and was sitting outside the
door, in the little courtyard at the side of the house. Sam,
tied up temporarily because he had been up to his tricks
again, was sitting on his haunches beside him, watching
Matthew career up and down the drive on his bicycle, while
Pippa and Anna played dolls on the wall. There had been no
call-out for well over a week. No foolish tourist had
teetered up the fell in high-heeled shoes and broken an
ankle. No walker had gone up alone and, after a last-minute
change of plan, decided to stay overnight in a distant valley,
unaware of the all-night search he had left behind him. All
was peace. Tina had decided to cook a special curry. They
might even get around to eating it without being interrupted.

The phone rang. John answered it and then came into the
kitchen, and sniffed the aroma of curry regretfully.

'It's a Panel search. A couple of lads lost up on the fell.
Apparently their father left them to make their way to
Angle Tarn, and they never arrived. The local team have
been searching the area. Now they've called us out. I'd
better go.'

Tina turned the curry down low and reached for the
sliced loaf. This could be a long night. John would need
some sandwiches.

They were on their way within ten minutes, Sam alert
and barking with excitement, going mad in the back of the
car. Tina waved them goodbye.

'Put my curry in the fridge. I'll have it tomorrow!' John shouted out of the window as they sped off down the drive. Tina blew a kiss, and began the evening struggle to gather the children for bed. It was almost dark. The first stars were freckling the sky above the fells. As she was drying Matthew after his bath she looked out of the window, down over the valley to the fells which lay in ever-darkening silhouette against the sky. What would it be like to have two of your children lost . . . up there? Even now, it was so wild. She had climbed there. She knew it was beautiful. She was not afraid. But to be lost in that great silent vastness, to have children lost, somewhere . . .

She shivered and drew the children close to her. Matthew began to clamour for a story and the others followed. Hers were safe. Silently she wished John luck, as she had so often before. She was glad there were people like him, ready to go up and search. Even if the curry might be spoiled . . . it might be her children one day, up there . . . she could not protect them for ever. Please God it never would be . . .

It was summer dark when he got to the base of the fells. Here, in the west, it was luminous still, but cold enough, this high up. The local team had already searched the immediate area, but had then called out a Panel search. Several teams would now be on their way. As he pulled up at the side of the road, he recognised several cars. This was to be the rendezvous for their own team.

The boys had apparently been lost while climbing above the lake, close to Angle Tarn. Now it was full dark and there was no sign of them. Everyone was getting worried. The team who were co-ordinating the search had designated Langenhow team to search the western fell, leading up to the tarn, while John was asked to take the next valley by himself, and meet up later with the rest of the team. He set off.

It was amazing how soon you were alone, how soon the

lights fell back into the valley and the noise faded away. Considering they were only walking. Now, in this summer dark, the darkness itself was a friend, even though there was a chill in the air. But for anyone lost on the fell, maybe exhausted and frightened, perhaps even injured, the dark even now could be full of dangers. Young kids, stumbling about in the dark . . . it would be so easy for one of them to go over a crag, tired and confused.

John had taken the walker's path on the western side of the beck and Sam was running ahead of him up the steep-sided valley . . . The only sound was that rushing bubbling exuberance of water which was always there in the fells. It, too, was a friend. But otherwise you were alone. And, if you were afraid, if you did not know what you were doing, you could do stupid things—get disorientated, fall off crags, anything. But if you faced it, if you were not afraid, if you kept your head, you might be OK. So long as you were well-equipped.

There was Sam's coolight, gleaming greenly ahead of him. The comforting 'squelch' of the radio. They searched on, hour after hour. It was a narrow valley. Sam, working on the western side, was covering with air scent the whole of the valley to the east, as the wind sank and eddied down from the eastern flank. Then, as the valley widened out towards the dale head, John sent Sam first up the western flank, working backwards towards him into the wind, and then away over the thin waterfall which the stream had become at this height, far away up onto the eastern ridge.

At intervals he picked up the backchat on the radio. No one was having any luck. A helicopter was on its way from RAF Boulmer to help in the search. Everybody was getting anxious.

Just before dawn, when the eastern ridge had become visible against a faintly lightening sky, John stopped for a break. It was the coldest hour of the night, just before sunrise, and they were at well over two thousand feet.

Tina's hot coffee and sandwiches were very welcome. Sam came back out of the deep dark under the ridge and helped him finish off the last of the ham and half a Mars bar. Above them and now just visible in the growing light, was the huge bulk of the Knott, the summit where they had agreed to rendezvous with the rest of the team. John had been out of direct radio contact for some time, although he was picking up odd bits of cross-talk, because the team were down in the next valley, still searching; but as he climbed up to the Knott, he was able to talk to them as they climbed towards him.

The sun began to mount towards the eastern ridge. There were few birds at this height to make a dawn chorus, but the sunrise was none the less glorious for that. Light flooded the sky above the fell. Down in the valley below it was still night, but here, at the summit, the air was pure gold.

Sam came bounding back, barking for pleasure—perhaps because, after a long night, he could at last see where he was going. A black bird, perhaps one of the rare ravens, crarked, and flapped its way up into the pink sky, disturbed by Sam's presence. It was like the beginning of the world, clean and new. And for all of them there would be a sudden renewal of energy, despite the fact of having searched all night.

One by one the other team members came up out of the darkness from the next valley, and there were familiar voices around him on the summit. No one had had any luck. Everyone was tired, but there was new hope now with the dawn coming. Things would be so much easier now, if only the lads had had the sense to stay still, and were not lying injured somewhere, needing urgent medical help. It was cheering that the weather had not clamped down. Despite the cold, as long as the weather held off, the lads had a good chance.

From here they were able to contact Control. They were

asked to make their way along the footpath towards Angle Tarn, where all the teams were to rendezvous. Together they set off, still glad of their torches to make sure of the way on the steep path. Sam was running ahead, but every once in a while he would come back, his cold nose thrust into John's palm, checking that all was well.

'I think I'll just take a walk over beyond the ridge with Sam,' John said suddenly. 'There might be something there. You never know. And after all, the next valley on from here was where the boys were last seen. It might be worth a try. We'll catch you up later.' He set off, one other team member, another John, having volunteered to keep him company.

Then, just as he had begun to leave the ridge and plunge back into the deep night shadow of the valley, Sam was beside him, indicating excitedly. Hesitantly, their eyes still struggling to adjust after the glare of the dawn, they made their way down the fellside, Sam running backwards and forwards to lead them in. Suddenly, they saw a tent ahead of them in a hollow, hardly visible in the deep dusk.

'Funny,' said John. 'In the briefing no one said anything about a tent. But we had better check it out.'

As they got closer, shining their torches, they saw that it was a small two-man tent, zipped up and seemingly without life. Sam was standing to attention outside. Even in the pre-dawn greyness, John could see he was looking very pleased with himself. John knocked on the roof.

'Anyone at home. Hello! Hello!' Sleepy noises came from within, and after a few seconds the tent walls bulged and heaved as someone struggled to unzip the opening. A head poked out, to be promptly licked by Sam. Someone shone his torch.

'Excuse me. We're searching for two lost boys . . .'

'You don't have to tell me,' the voice said wearily. 'I came up here to get away from everyone and get some peace, and all I've had the whole night long has been folk knocking on my tent roof asking me if I am lost. Well, I'm not lost and I would like to get a wink of sleep before morning. Next time I think I'll camp in middle of motorway. I'd probably get a sight more peace!'

John suddenly recognised the voice of a friend from the village back in Garth How. Now he remembered him talking the previous weekend about an expedition to get a bit of much-needed relaxation! It was a bit rough. He just hoped his neighbour would see the funny side after he'd had a bit of sleep, but as if to emphasise the unlikelihood of such a hope, a helicopter appeared in the luminous sky above them, the clatter of its rotor blades making an earsplitting noise across the valley. They watched it as it fell away, its searchlight cutting a swathe through the blackness below.

'Sorry I couldn't be of any help,' a voice said behind them. 'I do hope you find the lads. But please . . . could I now go back to sleep?'

John turned round to offer his apologies for disturbing the long-suffering camper. He knew this valley had been initially sweep-searched. He had probably been woken up a dozen times during the night, by well-intentioned and conscientious searchers.

Then, with a sudden cold feeling, he realised that Sam had gone. He had not sent him away, but all at once he was not there. He shone his torch round him and whistled, but there was no corresponding bark or a reassuringly warm body flinging itself at him out of the night. Sam was still wearing his jacket with the coolight. But there was no talltale greenish glow anywhere in the dimness of the valley.

He made his way over to the other team member who was relacing his boot.

'Did you see Sam take off? Is he with you? I can't make it out. It's not like him . . .'

His companion shook his head.

'Haven't seen him since we first woke that poor bloke in the tent. Has he gone off, then? Don't worry. He'll be back.'

But John was uneasy. It was not like Sam to go off on his own, without being sent away. He realised with a rush how tired he was. And here, in this enclosed valley, it was much colder than on the ridge. He began to wish he had saved a little hot coffee for later, instead of drinking the last just before dawn.

'I think we had better go back up a little way. Let's cross to the ridge on the far side. He's probably following a theory of his own somewhere up there.'

Together they climbed once more up towards the ridge, still in deep shadow. The arête was silhouetted sharp and black against a golden sky. To look at it left them dazzled, so that looking back at where they were walking, they could see only its image blazing before them. They tried not to look. Then the sun popped over the ridge, exactly like a child's sun from a pop-up story book, and suddenly the valley was on fire. Sam was nowhere to be seen.

John quickened his pace and, shading his eyes, tried to scan the horizon.

'Here, Sam. Here, boy! I can't see a thing in this glare.'

There was a loud barking above him, coming from just below the ridge, echoing round the high crag.

'He's right up there. What's he up to?' John pushed his way up, spurred on by the urgency of the barking, the other team member a little way behind.

The barking grew closer, and suddenly Sam appeared right in front of them, bursting out of the bracken at the foot of the crag, running for John.

'He's found something,' said his companion. 'Does he always do that chomping thing? Never seen anything like it.'

Sam hurled himself at his master and placed his paws firmly on his chest.

'Good dog!' John fussed him. 'You're doing well today.

Just as long as you haven't found anyone else who is trying to get away from it all!'

With Sam running backwards and forwards excitedly, they broke off from the path and climbed the steep, bracken-covered scree towards the mass of crag which hung over the valley. Sam bounded up the side and disappeared. They could hear him barking.

'Whatever it is, it's up on the crag,' said John. 'We'd better hurry.'

They had both been climbing all night, and suddenly John realised how his legs ached. But Sam barked again and, spurred on, they increased speed up the steep cragside, scrabbling for footholds.

'If it's the lads, how on earth did they get here in the first place?' the other John panted before he ran out of breath. Their fingers felt for the top of the crag and they hauled themselves up.

It was full morning up here, a small plateau leading more gently up to the tops. On the far side were two sleeping bags, sheltering under an overhang, and there were two tousled heads and two pairs of hands making the most enormous fuss of Sam, who was loving every minute of it.

John grinned at his companion, all tiredness gone.

'I think we've found them! I'll get on the radio. Let's see if we can raise someone. And let their father know.' He picked up his set, while the other John handed out emergency rations.

The boys were in good shape, all traces of fear having vanished with the thrill of being found by a dog, especially one as soft as Sam. They were still snuggled in their sleeping bags, now augmented by an MRT bivvy sack, to keep loss of body heat to a minimum. It was still early in the morning and, at this height, very cold, and a sharp little wind which had blown up with the dawn was buffeting them on the crag.

It was a long way down, and the boys had been lost all

night, even though the weather had been merciful. It was important that they were got off the fell as quickly as possible. But John had been trying to raise Control without success . . .

'Control from Dog John . . . Control from Dog John . . . it's no good. They're over beyond the next fell. I can't get anyone . . .' There was a sudden, excited cry from one of the boys. A helicopter, a small shape against the sparkling lake which lay below, was coming back up the valley. It got closer and closer, but at the last moment, while it was still a long way below them, it began to veer off to the right as though it was going to traverse the western edge of the fell.

John pressed the button again. 'Let's hope I can remember the helicopter call-sign . . . Rescue 31 from Dog John. Rescue 31 from Dog John . . . we are below you on the fell. Have found the two missing boys. Request assistance. Over . . .'

Afterwards, the boys could not decide which part of being rescued was the most exciting. First of all, after a long and rather frightening night, when they began to feel they were the only human beings within a hundred miles, to be found by a delighted Search Dog, who appeared out of the blue and woke them up by licking their faces, was a pretty good experience. But then to be winched high above the valley and the lake into a helicopter and taken down to safety, as their rescuers stood below them and the distant lake seemed to swing round and round and round . . . that would be something special to tell their mates back at school. Then being interviewed on Radio Cumbria was not bad, and would ensure instant fame for a satisfying length of time . . . But under all the bravado, it was dead good to be home, safe and warm. The younger one remembered that Sam had barked a last goodbye as they were winched away. He was the best. He was their friend . . . perhaps he had even saved their lives. They would never forget him.

John, almost too tired to drive home, had not switched on the radio, but everyone else seemed to have heard that the two boys had been found by Search Dog Sam and were safe and well . . . he was just glad they were OK. The relief and the release from tension were taking their toll. He would snatch a couple of hours' sleep. Then he would get in to work. Sam slept and snored on the back seat, unaware that he was a star for a day. To him it was all part of the job . . .

18

Late in October the first snows fell on the upper slopes and in the very highest of the valleys. The trees were laden with berries. Local people said that it would be the hardest winter for years. Langenhow MRT paid tribute in its annual report to Sam and his work for the team, which had increased their search capacity enormously. John was still an 'off-comer', teased for his different accent, and his odd ideas. But now it was something to be enjoyed. He would play up to it to irritate Bill, who was as chauvinistic a Cumbrian as you could find anywhere. Deep down, John knew they had been accepted; were even regarded with a degree of affection. They had proved their worth.

That November, they had heavy snow all over Cumbria and the North. The training weekend was held in near-blizzard conditions with drifts building up in the gullies, reaching more than six feet deep. It was a good opportunity for avalanche training. John volunteered to 'body', which as usual he found an unnerving experience. He reckoned that anyone trapped in a snow chamber for more than an hour would begin to hallucinate. He was not reassured to hear a shout from above after ten minutes to the effect that a lone skier, hurtling down the mountainside, had just passed within two feet of his chamber. The poor man must be wondering why, as he shot across the snowfield, he was greeted by a shout of 'Hey, watch out! There's a body under there.' It must have spoiled his day.

John had only just got into bed on the Sunday night, and fallen into an exhausted sleep (just because you had a graded Search Dog it did not mean they did not work you hard) when he was woken by the phone ringing. It was SARDA's training officer, who had himself only just struggled back to his home across the Pennines, in roads partly blocked by snow.

He had had a request from Yorkshire police for as many dogs as possible to rendezvous on the moors above Fylingdales. John, resignedly, staggered downstairs in search of dry socks, to find Tina already down, making up a flask.

'What would I do without you?' He put his arms round her briefly. 'God, I'm tired. I ache in every bone.'

She kissed him. 'You'll be all right when you get going. Here, have this.' She handed him a cup of black coffee. 'What is it this time?'

'Four venture scouts, lost on the moor around Fylingdales. They lost contact with the main party late yesterday afternoon . . . I do mean yesterday . . . ?' He looked at his watch. 'It's going to be a long night. I'd better get going.'

His boots were soaked, but he would not notice quite so much with a couple of pairs of semi-dry socks inside. He made pretty good time, mostly down the motorway, with Sam asleep on the front passenger seat beside him. Once off the main roads, and driving across the moor, however, it was a different story, with the wind whipping up drifts feet deep on the exposed moorland roads. But the by-roads were being watched by police and RAC/AA men, and John had a kindly escorting Land-Rover for the last few miles.

They searched all night, in the bleakest conditions John could ever remember. At 7.12 a.m. the boys were found, bivvying in a sheltered position, snug in their sleeping bags behind a protective wall of snow. They had survived one of the worst nights the moor had ever known . . . at least this century. A snow plough led the little procession of SARDA handlers, police and ambulance back to the main road. They were all buoyed up: by the fact that a dog—Ruff and his handler Joe Broughton—had 'found', which was always good for SARDA, and because they had pulled four live boys off the moor, suffering from very little except mild exposure, when they had expected, with almost 100 per cent certainty, to find four frozen little bodies. As the team leader said later:

'If only more youngsters were sent out with the right equipment and intelligent instruction, how many more lives would be saved!'

John went in to work for the last two hours of the afternoon, aching all over, but strangely, no longer tired. Sam, however, had not even bothered to open his eyes as John let himself out of the house. He was flat out.

But as soon as John got home in the evening, it hit him, and he passed out in an armchair in front of the fire, with his supper on his knee. Wisely, Tina left him alone. But at ten-thirty that night, just as Tina was steeling herself to wake him and get him up to bed, the phone rang again. It was an apologetic Bill Nicholson.

'Lass, I'm sorry. We've had another call-out. It's allus the same. One after t'other. We've all been out three times in last forty-eight hours . . .'

'He's whacked, Bill. Isn't there anyone else?'

'Have tha seen t'wedder? Roads are blocked . . . there's nae more than three dogs ower this side not injured or snowed oop on their farms. And it's a serious one, lass. Two climbers lost out on fell, in this! Ower at Muddlewater. They'll not survive unless we find 'em quick . . .'

'I'll get him . . .' she said resignedly, only to find John's hand reaching from behind her to take the receiver . . . he smiled at her and winked . . .

'Thanks for trying! What's the score, Bill . . . ? I'm on my way.'

They made it to the rendezvous at the edge of Muddlewater in twenty-five minutes—not bad going considering the conditions. The car park serving the

Nature Reserve hut (closed for the winter) was jammed with cars, police Land-Rovers and an ambulance. There were flashing blue and orange lights, headlights of cars, torches, between them turning the snowy banks around into a fairyland of colour.

'Big rescue, Sam. You'll have to be on your best behaviour,' John said as he let him out of the car and fitted on his jacket. He struggled into his boots, duvet jacket, gaiters, waterproof trousers and balaclava. Even then he could feel the wind tearing through him. The snow was coming at them straight across the lake with no shelter. He had to fight to shut the car doors.

They battled their way over to the Control Land-Rover, manned by Scarpdale MRT. The Controller explained that they had been searching Scarpdale, the fell above Muddlewater, since four o'clock that afternoon, after a climber had rung the local police from a hostel in the valley saying that he had become separated from his friends during a fell walk, when the weather had suddenly deteriorated.

At best, they were searching for two benighted walkers, well-equipped and experienced (if their companion was to be believed), who had probably 'dug in', and could be found in a limited area. At worst, two men who might be seriously injured, lost anywhere in a vast area of fell, who, however well-equipped (though not unfortunately with a tent), would certainly die of exposure if they were not found within a certain number of hours.

John and Sam were assigned an area about a thousand yards square on the distant edge of the lake, well beyond the path which the walkers were supposed to be following, but nonetheless an area which must be 'cleared'. It was a part not yet searched by men on the ground, for which John was grateful. It always made things more difficult if the area had already been walked over.

He examined the map. It was tricky terrain, the sort of

ground which would take twenty men several hours to search. It was full of gullies and crags and granite outcrops rearing out of the fellside. The contour lines looked like the surface of the moon, with whorls and indentations showing the sharp changes in height.

But it was ideal territory for a Search Dog. Properly employed, a dog would cover the area thoroughly in two or three hours, even in these near white-out conditions.

They fought their way up the fell, leaving the blazing lights of the car park to the right, temporarily blinded by the completeness of the dark. At night there was always that sense of plunging into some great, all-embracing loneliness, of cutting away the rest of the world, of taking a deep breath and leaving familiar things behind. Particularly on a night like this. Then they would settle down: the comfortable 'squelch' of the radio would begin; the sudden bursts of cross-talk; Sam would go off and return. It was routine work, requiring intense concentration, but in the vast solitude of the night, you were glad of it.

They crossed the path which the walkers were supposed to have used. It was invisible now, but John, holding a detailed map of the whole area in his head, had little difficulty in recognising it. They climbed on and on into the fastness of Scarpdale. At one point John stopped for a breather, and looked back. It was no longer snowing and a gibbous moon was hanging in the sky above Muddlewater. The fells around the lake were lit with a silver light and in the centre, like a glittering jewel in a crown, moonlight shimmered on the sleeping lake.

Sam came bounding back and flopped down beside John, sending up a little flurry of powder snow in the torchlight. John put his hand down and stroked the smooth skull.

'Look at that, Sam. Isn't it the most beautiful sight you ever saw? No wonder people are drawn to the fells; even take risks to get up here. There's some sort of mystery . . . it's like going back to the beginning . . .'

There, immediately below them, looking strangely incongruous, was the blazing square of the car park, which was control HQ for the Search and Rescue. It was alive with moving lights and seemed very far away. Only the silence, and the wind, and the great fells, seemed truly alive.

As though to punish him for his romantic notions, the snow hit him ferociously once again. Sam, the car park, the lake, everything was blotted out with terrifying suddenness. John put his hand down to feel for Sam's collar; to keep him with him until he could get his bearings. But there was nothing there. He groped again, with a sense of sudden, unreasoning panic. Then, to his relief, he felt Sam's body against his knee on the other side—he had become completely disorientated. Sam licked his hand, reassuringly. It was good to have him there.

They walked on for an hour or more. The snow eased off. John carried on up the fell, and sent Sam obliquely into the wind. It was hard work, walking. Because of the wind there was hardly more than a dusting of snow in some places; in others, sudden, unexpected drifts which caught the walker unawares. And underneath, the ground was frozen solid; lethally slippery.

Suddenly, John heard Sam barking up ahead, and at the same moment the last of the snow flurries left him behind; only the wind now snatched up the snow in thin trails of spindrift across his path. He heard the barking again, and as the moonlight flared once more over the lake, he saw Sam silhouetted in an instant on an outcrop of rock which seemed to overhang the water. Sam was running round in an odd way, very excited. Then he vanished.

John was suddenly sick at heart. He began, clumsily, to pound over the frozen bracken, his own breathing loud in his ears. In a minute, Sam would come back to him. He always came back. First he indicated, then he came right up to John—to make sure. John could see a clear white snowfield ahead of him, lit by the moon. But there was nothing.

He ran on, sweating now with horror. It could not be that Sam had gone over the edge of the crag. It could not. He was excited. He had probably 'found'—perhaps on the far side of the granite outcrop. Any minute he would come bounding down, in that beloved way, his jaws chomping together with excitement.

Every muscle screaming with effort, he flung himself up the side of the overhang which, from the north, was nothing more than a steep slope like a ski-jump. He raced to the top, calling Sam's name over and over again.

It was a sheer drop. Underneath lay, not the lake, but a long, snow-covered incline sweeping down to the shore. And, directly underneath the cliff where the snow lay white and smooth—forty feet below—there was a small, dark shape.

John never knew how he got down those forty feet. It seemed to take forever, and it seemed to take no time at all. But now he was there, the snow coming up to the top of his thighs, half-blinded by the drift of it. And Sam lay spreadeagled on the whiteness, strangely still, a creeping stain spreading outwards from his head.

John knelt down and picked him up in his arms. He began to run up the killing slope, somehow finding his footing, up the gullyside behind the brooding overhang, cut off from the moon—sobbing for breath; his arms dead already from the weight of Sam. And Sam's head cradled in his arms—matted with blood, the great incisors exposed and gleaming. Only the faint heartbeat and the terrible, snuffled breathing telling John that he was still alive . . .

*   *   *

They tried to contact John, but there was nothing. The two climbers had been found—dug into a snow chamber on the farthest western extremity of the fell; riding out the storm.

Now, anxious SARDA members and their dogs, concerned that the weather was worsening once again, set off up the fell to the farthest point above the lake, which was John's assigned 'patch'.

A few yards up the fell path, as they struggled through the blinding snow, Joe and Ruff, the leading pair, were met by a desperately running figure in the last stages of exhaustion, carrying Sam in his arms. They were both covered in blood. At first John refused to let Sam go, so a stretcher was rushed up from the car park and in the end accommodated them both. And by the time they got down again, to the waiting ambulance, John was too far gone to know that Sam had been taken away.

*   *   *

That should have been the end. But this is a true story, and Sam somehow, miraculously, survived. Above one eye he bears a deep scar where the flesh was bared to the bone and his eyelid no longer closes properly, giving him a raffish look. The vet, pulled out of bed by a Land-Rover sounding a klaxon beneath his window, worked all night to save him. And just about dawn, after a blood transfusion and a drip to counteract the deep shock, and twenty stitches, Sam began to fight back.

As for John, his SARDA colleagues say that they were more worried about him than about Sam when he hurled himself at them out of the blizzard. That may be an exaggeration, but he was in a pretty bad state—stretched beyond the limits of exhaustion by superhuman efforts; deep in shock and distress. And the scars are still there: a permanently bad back, a torn shoulder which aches on wet days, memories which return in dreams of that horrific run down the fell with seventy-five pounds of dog in his arms and Sam's blood running down into the snow.

But one question still puzzled him . . . why had Sam behaved so oddly and become so excited on the top of the crag, when the 'bodies' were over a mile away at the farthest end of the lake?

The following spring, they went up alone together, retracing the terrible journey of the previous winter, to the place where Sam's blood had stained the snow. As they walked, John remembered that other spring, when they had quartered this very ground for the man from the fellside cottage, before the snow had stopped the search. He had never been found. The fells hid so many secrets . . . they had been there such a long time, and had long memories.

John could not bear the thought of climbing onto the overhang; instead they worked their way up from lower down, where the fading snow revealed a path among the precipitous rocks. Sam seemed to have no memory of the place, but ran ahead gaily. John looked up. The fall had been forty feet or more. Sam had been so terribly lucky. He had not broken a single bone in his body, although he was appallingly bruised. If there had not been deep snow . . .

Suddenly, John was jerked out of his reverie. Sam was barking ahead of him. For an instant, the whole of that dreadful night flashed in a series of pictures across his mind. Then he shook it off. Sam was racing towards him through the snow, jaws chomping with excitement.

'Show me, Sam! Show me!'

He picked his way carefully through the half-frozen snow, mindful of how teacherously unstable the whole rock face would be, now that the snows had melted. Sam came back once, twice, and eventually led him, past the spot where he had fallen, to a corner of a gully.

There was a skeleton laid out among the rocks, hidden from above and below. Some of the bones were broken, as though he or she had fallen from the overhanging crag above. It was onto this merciless bed of scree that Sam, too, had fallen, but there had been snow to cushion him and break his fall.

John moved forward. The foxes and the crows and

216

numerous snowfalls had done their work. There was very little left. How long had the body lain here, undiscovered? And how had Sam ever picked up a scent from this, under feet of snow? Beside the skeleton was the ancient frame of a haversack, a few tattered remnants of cloth still adhering to the struts and, strangely, still held by the fragile weight of the body, some shreds of whitened plastic, bleached by the weather.

John stood up. Sam, who had been sitting at his feet, whining softly in his own slow distress, understanding everything, bounded off up the crag, reappearing at the top. John's heart gave another lurch . . . but this time Sam stood foursquare, firm and strong, looking over the lake. If he remembered, he was no longer afraid.

John climbed up beside him and together they looked out over the stillness of Muddlewater. From up here, the lake was deep azure, seeming unruffled by wind, reflecting back the golds and browns of old bracken on the far side. Behind them crouched the great bulk of Tarn Howe, like a great moulting animal, with its clumped firs. From far below, a ceaseless chorus of young hoggets calling for their mothers drifted up at them from the intake fields. A buzzard wheeled away on the wind.

But suddenly, John found he was staring at something just by his feet. A clump of daffodils had taken to the soil on the green knoll at the head of the crag. They were framed, bright and free, against the blue lake. A perfect place. His eye was drawn sideways again . . . back towards Tarn Howe. And there he saw, irregularly planted, clumped on either side of the path, the hundreds and hundreds of daffodils which the spring had brought out of hiding.

And suddenly, the last piece of the jigsaw fell into place. He raised his head. The snowy peaks on the far side of the lake were caught once more with gold. He smelt the cold snow wind. The daffodils danced.

Jed Cross had planted a fine memorial.

217

Epilogue

Sam is nine now. His hard life has left him permanently out-at-elbow on one side, from a sprain he had a while ago. The scar above his eye marks forever his fall from the crag. His nose looks as though he has been digging again. He will never be beautiful.

And now there is Tyan (Cumbrian for Two), a Labrador cross puppy who will eventually succeed Sam as a Search Dog. He has an abundance of enthusiasm, following Sam around on his practice finds and racing back to indicate to John before Sam can do it himself. Sam does not at all like being upstaged. But that day, when Sam has to retire, a day which seemed so far away when I began writing this book, is now sadly very close . . .

This February I travelled to Cumbria for the four-day Annual Course, sponsored for the second year by Pal, to watch Sam being assessed for regrading for the last time. We were all a little tense. Sam was going mad with excitement whenever he thought a search was in the air. The suspense was made worse by having to wait until the Sunday for the regrading assessment. John was fully committed to examining the trainee and Novice dogs, perched high on a hill under Blencathra in the freezing wind, with Sam by his side. But at last, on Sunday lunchtime, after I had done a spot of bodying myself, eating my sandwiches under a waterfall, two bodies were put in, high up on the Mosedale valley, and Sam was given an hour to find them. They set off down the track while I stayed in the Land-Rover with the assessor from SARDA Wales.

I kept thinking of this book and how I would end it if Sam was too stiff to make the grade. It was by no means automatic that he would pass. SARDA cannot afford to have unfit dogs in the team. They are a weak link, and there is no place for sentiment. And I knew Sam's shoulder had been giving him some trouble of late, although once there was a call-out, all stiffness seemed to be forgotten. I waited anxiously.

Suddenly John appeared high on the skyline, far up above the scree, and simultaneously the tiny yellow shape of Sam could be seen streaking away diagonally across the dark fell. My heart lifted. I had watched enough dogs by now to see how, in one fell swoop, Sam, by working across the wind, had eliminated a huge area of bracken, boulders, and loose scree, which would have taken a line search hours to cover. That was what it was all about. We watched Sam as he worked up and down. Through the glasses I could see a madly wagging tail, and John keeping to the skyline, working Sam with occasional hand signals. Even from this distance we could all feel the accord between them. It was like some strange balletic movement . . . perfectly choreographed. Someone beside me, watching, said:

'It's not often you get a chance to watch these old dogs going through their paces. It's something to see!' I was taken aback. Tears pricked behind my eyes for a moment. Sam . . . an old dog! Only such a little time ago he had been a beginner, getting into trouble, disgracing himself, behaving like an adolescent. Sam . . . a senior citizen? It was hard to believe.

But Sam still had a surprise up his sleeve. Everyone knew that, whatever happened, he would never come up for regrading again: this was his swan song. Suddenly, across the fell, just audible above the rushing waterfalls in the valley below, came the clear sound of an indication. The assessor grabbed the glasses and watched intently.

219

'What's he playing at?' he said after a few moments. My heart sank. Surely, after that wonderful performance, Sam was not going to blow it now.

'What's happened?' I asked.

'He's come back to John and indicated. He seems to have begun leading John in. That's definitely where one of the bodies is hidden. But now he's gone off in quite another direction. I can't make it out.'

We watched as the small figure bounded up the fell once more. Then, after a few moments, on the clear air, we again heard Sam's distinctive bark.

'Well, he's got it right. That looks like both bodies found.'

I got out of the Land-Rover and began to make my way down the track. It was clear and cold, my cheeks burning in the stinging wind. I thought it was going to be all right. I saw John approaching through the bracken. Sam bounded up to me and snuffled for chocolate bars. I gave him a half one which I had saved from my lunch.

'Trust you, Sam. Trust you to come up trumps.' He pushed his nose into my hand by way of farewell and then galloped on down the path. There might be other packed lunches to raid, with any luck. John followed more slowly.

'Did you see it? I hope they realised what was going on. It took me a few minutes to realise myself,' he said when he came up to me. 'The first body said he went in and said hello. Then he did a U-turn and went straight up to the second and led me right in. I couldn't believe what was happening. Could it be two bodies—in one sweep? But I did take him back to the area where he had first indicated, just to make sure, and he led me straight to the other body, who must have been lying there, wondering when Sam was coming back.' He shook his head. 'Two bodies in one sweep. Trust Sam.'

'He looked marvellous,' I said. 'I was proud of him.'

Later that day we heard that Sam had passed his regrading with flying colours. He will probably go on for another year, and then John will retire him with honours. He will have done enough.

Then life for Sam will become a little more peaceful. He will probably spend his time sitting in his favourite spot outside the door, among the geranium pots and children's toys. And when the telephone rings, and the call-out comes, he will no longer rush excitedly to the car, yelping with anticipation. Soon he will have to learn that it is no longer for him. It is hard to say goodbye; it is hard to leave the race behind, to see other dogs taken off into the night, scarlet-jacketed, ready to be heroes.

Yet there is another year at least. On the last day of the course, at one o'clock in the morning, when some were sleeping and others involved in a rousing sing-song, the Search Dogs were called out by Keswick Team to search for a young lad overdue at Buttermere. Sam, newly regraded, spent the night with John on the fell, including a brief sleep in a bivvy bag together just before dawn. It was bitterly cold and it was not until after first light that the boy was found, alive and well.

The work goes on. Sam is only one of the increasing numbers of dogs and handlers helping to save lives somewhere, almost every week of the year. Each one is part of an MRT effort. Sam has only been a tiny element of a magnificent Search and Rescue organisation which stretches from the northernmost tip of Scotland to the slopes of Snowdon and the wilds of Dartmoor, and across the Irish Sea. Its influence has been felt as far away as the crushed and teetering destruction of San Salvador. And it may be that the El Salvador Experience is only a beginning for International Search and Rescue.

But for us, who have come to know Sam well, and for whom he will always be special, as each creature should be loved and particular to his own hearth and those who know him well, it will be very hard to say goodbye . . .

This is SARDA's book. It is for all the men and women,

and for every dog who has ever worked out on the bare fell and the empty moor and the high crag, seeking human scent on the wind. It is also, I must confess it, for my own dogs, one of whom, my first black Labrador, Bess, I loved and lost most tragically. But most of all, it is a book for Sam alone. For all he has given; for the generosity of his spirit; for his loyalty.

Simply with love.

Donations to SARDA, which is a registered charity, should be sent to:

The Hon Treasurer
SARDA
Greenstones
Old Hall Road
Troutbeck Bridge
Windermere
Cumbria